T0390554

VOLUME SEVENTY FIVE

ADVANCES IN CATALYSIS

Computational Insights into Catalytic Transformations

VOLUME SEVENTY FIVE

ADVANCES IN CATALYSIS

Computational Insights into Catalytic Transformations

Edited by

MARIA BIOSCA

Departament de Química Física i Inorgànica,
Universitat Rovira i Virgili, Tarragona, Spain

Academic Press is an imprint of Elsevier
125 London Wall, London, EC2Y 5AS, United Kingdom
50 Hampshire Street, 5th Floor, Cambridge, MA 02139, United States
525 B Street, Suite 1650, San Diego, CA 92101, United States

First edition 2024

Notices

Knowledge and best practice in this field are constantly changing. As new research and experience broaden our understanding, changes in research methods, professional practices, or medical treatment may become necessary.

Practitioners and researchers must always rely on their own experience and knowledge in evaluating and using any information, methods, compounds, or experiments described herein. In using such information or methods they should be mindful of their own safety and the safety of others, including parties for whom they have a professional responsibility.

To the fullest extent of the law, neither the Publisher nor the authors, contributors, or editors, assume any liability for any injury and/or damage to persons or property as a matter of products liability, negligence or otherwise, or from any use or operation of any methods, products, instructions, or ideas contained in the material herein.

ISBN: 978-0-443-31334-9
ISSN: 0360-0564

For information on all Academic Press publications
visit our website at https://www.elsevier.com/books-and-journals

Publisher: Zoe Kruze
Acquisitions Editor: Mariana Kuhl
Editorial Project Manager: Palash Sharma
Production Project Manager: Sujatha Thirugnana Sambandam
Cover Designer: Gopalakrishnan Venkatraman

Typeset by MPS Limited, India

Contents

Contributors

Mercè Alemany-Chavarria
Departament de Química and Centro de Innovación en Química Avanzada (ORFEO-CINQA), Universitat Autònoma de Barcelona, Cerdanyola del Vallès, Catalonia, Spain

Maria Besora
Departament de Química Física i Inorgànica, Universitat Rovira i Virgili, Tarragona, Spain

Maria Biosca
Departament de Química Física i Inorgànica, Universitat Rovira i Virgili, Tarragona, Spain

Jorge J. Carbó
Departament de Química Física i Inorgañica, Universitat Rovira i Virgili, Tarragona, Spain

Montserrat Diéguez
Departament de Química Física i Inorgànica, Universitat Rovira i Virgili, Tarragona, Spain

Fahmi Himo
Department of Organic Chemistry, Arrhenius Laboratory, Stockholm University, Stockholm, Sweden

Feliu Maseras
Institute of Chemical Research of Catalonia (ICIQ), The Barcelona Institute of Science and Technology, Tarragona, Spain

Gantulga Norjmaa
Departament de Química and Centro de Innovación en Química Avanzada (ORFEO-CINQA), Universitat Autònoma de Barcelona, Cerdanyola del Vallès, Catalonia, Spain

Oscar Pàmies
Departament de Química Física i Inorgànica, Universitat Rovira i Virgili, Tarragona, Spain

Giuseppe Sciortino
Departament de Química and Centro de Innovación en Química Avanzada (ORFEO-CINQA), Universitat Autònoma de Barcelona, Cerdanyola del Vallès, Catalonia, Spain

Albert Solé-Daura
Institute of Chemical Research of Catalonia (ICIQ-CERCA), The Barcelona Institute of Science and Technology, Tarragona, Spain

Gregori Ujaque
Departament de Química and Centro de Innovación en Química Avanzada (ORFEO-CINQA), Universitat Autònoma de Barcelona, Cerdanyola del Vallès, Catalonia, Spain

Preface

Catalysis is a vibrant and fundamentally relevant field of research, which plays a pivotal role to address present and future challenges in synthesis. Many research areas and industries depend on catalysts to construct molecules that, in turn, can inhibit the progression of diseases, form elastic and durable materials or store energy in batteries among many other applications. The discovery of an efficient catalyst is mostly carried out empirically, ranging from trial-and-error approaches to more or less rational designs based on mechanistic studies. This process has been aided by a variety of procedures for evaluation of catalysts, such as high-throughput experimentation, but still remain costly. These semi-empirical approaches can reject good catalysts that unfortunately deemed not successful during the initial screening not because the catalyst was not good but because it was not tested in the right conditions. Nowadays catalyst discovery is increasing being aided by the advances in DFT computational simulations, both from the methodological (new functionals, better suited for the description of catalysts have been developed) and the practical (real systems can be now treated at full DFT level in a considerable reduced time) perspectives.

In this context, the volume "Computational Insights into Catalytic Transformations: Exploring the Potential of DFT Calculations" provides readers with a perspective on the transformative impact of computational chemistry on the understanding and development of catalytic processes. This volume of Advances in Catalysis comprises four chapters, each focusing on a distinct area within catalysis and authored by world-leading experts in their respective fields.

Chapter 1, authored by M. Biosca and F. Himo, delves into the complex mechanisms behind the synthesis of various benziodoxole reagents and the exploration of the reaction mechanisms and sources of regio- and enantioselectivity of electrophilic fluorine transfer reactions using in situ generated hypervalent iodine catalysts. Chapter 2, by M. Biosca, M. Besora, F. Maseras, O. Pàmies, and M. Diéguez, presents several examples where DFT calculations are employed to elucidate the factors influencing enantioselectivity in two metal-catalyzed asymmetric transformations that were crucial to find the optimal catalyst. Chapter 3, contributed by M. Alemany, G. Norjmaa, G. Sciortino, and G. Ujaque, offers an overview of chemical processes catalyzed by Metal-Organic Complexes (MOCs) from a computational perspective. This includes studies providing a molecular description of the binding processes and selected examples of reactions that are accelerated by MOCs and have been examined computationally. Lastly, the chapter by A. Solé-Daura and J. J. Carbó provides a comprehensive review of the computational contributions to understanding olefin epoxidation by hydrogen peroxide catalyzed by substituted polyoxometalates.

This volume is intended for a diverse audience, including organic, catalytic, organometallic, computational, and industrial chemists. Our objective is to present a comprehensive overview of the principles and applications of DFT in catalysis. You will find case studies that underscore the practical benefits of computational insights, discuss the latest advancements in DFT methodologies, and provide guidance on the effective implementation of these techniques in research. Whether you are an experienced researcher seeking to deepen your understanding of catalytic mechanisms or a newcomer to the field desiring a robust introduction to computational methods, we hope this volume will serve as a valuable resource.

Finally, we extend our gratitude to Ms. Mariana L. Kuhl and Mr. Palash Sharma, whose assistance during the editing and production process was invaluable. We also acknowledge the support of FEDER/Ministerio de Ciencia e Innovación (MICINN)/AEI for grant PID2022-139996NB-I00 and for the Juan de la Cierva fellowship awarded to Maria Biosca, as well as the Catalan Government for grant 2021SGR00163. Finally, we would like to express our special thanks to all the authors who contributed to this project whose expertise and dedication have made this work possible.

Maria Biosca

CHAPTER ONE

Exploring reactivity of fluorine transfer hypervalent iodine reagents: A quantum chemical perspective

Maria Biosca[a,*] **and Fahmi Himo**[b,*]
[a]Departament de Química Física i Inorgànica, Universitat Rovira i Virgili, Tarragona, Spain
[b]Department of Organic Chemistry, Arrhenius Laboratory, Stockholm University, Stockholm, Sweden
*Corresponding authors. e-mail address: maria.biosca@urv.cat; fahmi.himo@su.se

Contents

Abstract

Hypervalent iodine compounds have gained widespread utility in modern organic chemistry. A number of benziodoxole derivatives have been synthesized and used as reagents to facilitate the transfer of fluorine or fluorine-containing substituents onto organic substrates. Also, novel reactions have emerged using in situ generated electrophilic hypervalent iodine species, utilizing HF as a fluorine source. Detailed knowledge about the reactivities and modes of action of these compounds is of great importance for the development of new reagents and catalytic protocols within this field. To that end, quantum chemical methods have contributed significantly to the elucidation of the mechanisms and the factors determining selectivities of this important class of reactions.

In this chapter, we present examples from our recent work using density functional theory calculations aimed at shedding light at various aspects of this chemistry. We focus on the intricate mechanisms for the syntheses of various benziodoxole reagents and on the understanding of the reaction mechanisms and sources of regio- and enantioselectivity for organo-catalyzed electrophilic fluorine transfer reactions employing in situ generated hypervalent iodine reagents. These studies have yielded

Advances in Catalysis, Volume 75
ISSN 0360-0564, https://doi.org/10.1016/bs.acat.2024.08.004

novel mechanistic insights with potential implications for other reactions involving the incorporation of fluorine or fluorine-containing groups with hypervalent iodine reagents/catalysts.

1. Introduction

Medium-sized molecules are essential for various societal needs, such as, for example, pharmaceuticals and agrochemicals. The incorporation of fluorine or fluorine-containing substituents into these molecules confer them unique properties, altering their physical and chemical characteristics, such as metabolic stability, lipophilicity, and membrane permeability *(1–3)*. Therefore, in recent years, the significance of compounds containing fluorine has increased considerably and, as a result, the development of new effective and selective methods for fluorinating organic molecules has become increasingly important *(4–8)*. Great advances have been made in the exploration of techniques that incorporate the fluorine atom into the carbon frameworks of chemically relevant target molecules via fluorination (R-F), trifluoromethylation (R-CF_3) or trifluoromethylthiolation (R-SCF_3) reactions. To this end, an important approach to effectively introduce fluorine or fluorine-containing substituents into organic molecules is through the use of hypervalent iodine transfer reagents. This approach enables a variety of synthetic transformations, including the fluorination of alkenes, the difunctionalitzation of diazocarbonyl compounds, and fluorocyclization reactions, among others *(9–14)*. The effectiveness of hypervalent iodine reagents in these transformations is due to their distinctive electrophilic nature. This unique reactivity arises from the presence of a 3-center-4-electron bond, so called hypervalent bond. This type of bond is longer, more polarized, and weaker than a typical covalent bond, resulting in increased electrophilic reactivity *(15–19)*. Typically, the incorporation of the fluorine atom or the fluorine-containing substituents from the hypervalent iodine into carbon-centered nucleophiles occurs through two methods. The first one involves the use of hypervalent iodine compounds as reagents. These reagents contain the fluorine or the fluorine-containing substituent (e.g. compounds **1–4**, Scheme 1) and are capable of releasing the essential "electrophilic fluorine or fluorine-containing substituent" to the organic molecule. The second approach involves the use of iodoarenes for in situ generation of the electrophilic hypervalent iodine species with an oxidant or the use of hypervalent iodine reagents such as phenyliodine(III) diacetate (PIDA, **5**) or [bis(trifluoroacetoxyiodo)]benzene (PIFA, **6**) along with the employment of a

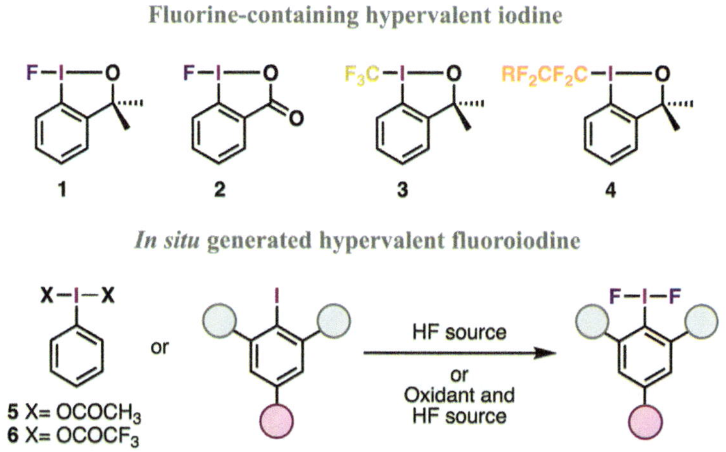

Fluorine-containing hypervalent iodine

1 2 3 4

In situ generated hypervalent fluoroiodine

X–I–X

or

HF source

or

Oxidant and HF source

F–I–F

5 X= $OCOCH_3$
6 X= $OCOCF_3$

Scheme 1 Examples of hypervalent iodine reagents/catalysts used for the incorporation of fluorine or fluorine-containing substituents into organic molecules.

fluorine source such as py·HF or NEt_3·HF (Scheme 1). The fluorination takes place from the in situ generated reactive fluoroiodine(III) intermediate. These methodologies not only advance fluorination chemistry but also provide avenues for tailored synthesis of fluorine-containing molecules with diverse functionalities and applications.

Many aspects of the of fluorination reactions involving fluorine-containing hypervalent iodine reagents, such as compounds **1–4**, have been investigated using both experimental *(20–23)* and computational methods *(24–32)*. These studies have led to a deeper understanding of the reactivity of these fluorine-containing benziodoxole-based reagents. Similarly, various features of the formation and the reactivity of in situ generated fluorine-containing hypervalent iodine catalysts have been elucidated *(33–39)*.

Detailed knowledge about these mechanisms offers perspectives for innovating new reactions and devising procedures to control and enhance the selectivity of the reactions, thereby advancing the frontier of organofluorine synthetic chemistry. Quantum chemical methods have become an essential tool in the elucidation of mechanisms of organic and organometallic reactions. In particular, Density Functional Theory (DFT) calculations are widely employed for this purpose.

In this chapter, we will review a number of examples from our recent work to illustrate the power of this technique to uncover very detailed reaction mechanisms, and thereby gaining deeper understanding of these processes. More specifically, we will discuss how the calculations were used to uncover the mechanisms for the syntheses of benziodoxole reagents *(40)*. We will also show how the reaction mechanisms and sources of regio- and enantioselectivity for organo-catalyzed electrophilic fluorine transfer reactions employing in situ generated hypervalent iodine reagents were elucidated with DFT calculations *(41,42)*. The developed insights are important for the advancement of novel and improved reagents or catalytic systems in this fascinating field of chemistry.

2. Computational details

In all the examples discussed below, the dispersion corrected B3LYP functional was used in the calculations *(43–46)*. Geometry optimizations were typically carried out with a medium-sized mixed basis set, comprising effective core potential for metals and I, and 6–31 G(d,p) for the other atoms. For enhanced accuracy, single-point calculations were performed on the basis of the optimized structures with the same basis set for metals and I, and with 6–311 +G(2d,2p) basis set for all other elements. Implicit solvation was included in geometry optimizations employing the SMD method *(47)* with the solvents used in the experiments. Frequencies were calculated at the level of theory of the geometry optimization to obtain Gibbs energy corrections. Standard state corrections were also added.

In all cases, a thorough conformational search was conducted for all intermediates and transitions states to ensure the identification of the conformation with the lowest energy. This procedure is very important, since the models are becoming larger and many rotamers and complexation geometries are possible *(48–52)*. This is especially critical for the study of enantioselectivity, such as in the case of the second example discussed below (Section 4), as this is dependent on reproducing very small energy differences.

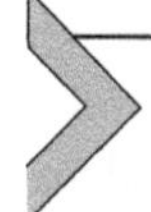

3. Formation of benziodoxole-based CF_3 and SCF_3 transfer reagents

The benziodoxole-based reagent **3** (called Togni reagent) stands as one of the most frequently employed benziodoxole-based reagents and it is well established that it exists in its hypervalent iodine form. **3** is straightforwardly synthetized from chloro-benziodoxole **7**, which is converted to fluoro-benziodoxole **1**. Following this, compound **1** reacts with the Ruppert–Prakash reagent **8** in the presence of potassium fluoride (Scheme 2) *(53)*. The trifluoromethylthio analogue of the reagent, on the other hand, exists in the thioperoxide form (**11**), and is synthetized from the reaction of chloro-benziodoxole **7** with $AgSCF_3$ **9** *(54,55)*. Thus, while **3** can be easily synthetized in the hypervalent iodine form, the corresponding **10** for the trifluoromethylthio reagent has not been observed.

The ether form of the reagents is in general much more stable than the hypervalent form *(56)*. Nevertheless, some benziodoxole-based reagents, such as **3**, remain kinetically stable, while others undergo rearrangement to attain the thermodynamically favored isomer *(56,57)*. The factors underlying these distinct behaviors were analyzed by examining the mechanisms for their syntheses by means of DFT calculations *(40)*.

The mechanism obtained from the DFT calculations for the synthesis of Togni reagent is shown in Scheme 3, and selected optimized structures are displayed in Fig. 1. First, a stable adduct formed between the KF salt and fluoro-benziodoxole **1** was identified by the calculations (**Int1** in Fig. 1). The reaction starts with the nucleophilic attack of the fluoride ion of KF (from an alternative adduct structure **Int1'**) onto the silicon center of the Ruppert–Prakash reagent **5** forming **Int2**. Next, a reorientation occurs to form **Int2'**, whereby the CF_3 group is directed towards the potassium ion. A heterolytic dissociation of the Si–CF_3 bond takes then place via **TS2** (Fig. 1) to release of the trimethylfluorosilane $(CH_3)_3SiF$ as side product and generate intermediate **Int3** (Scheme 3), comprising a CF_3^- group.

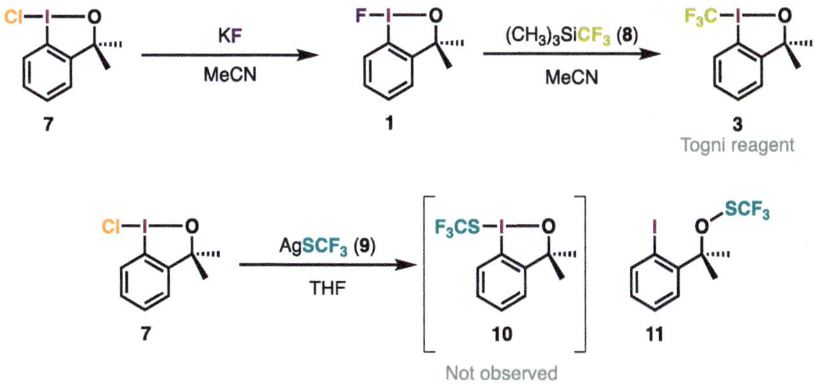

Scheme 2 Syntheses of the fluorine-containing hypervalent iodine reagents: (1) Togni reagent **3** and (2) benziodoxole-based SCF_3 transfer reagent **11**.

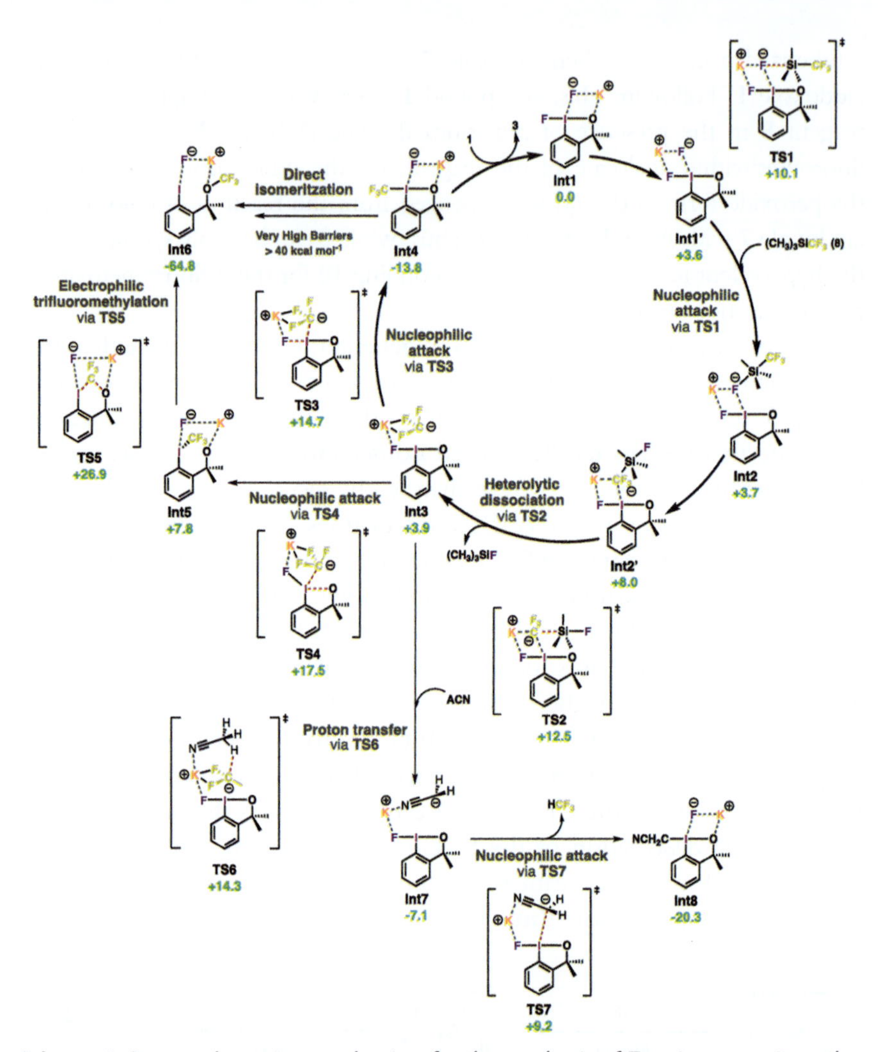

Scheme 3 Proposed reaction mechanism for the synthesis of Togni reagent **3** on the basis of DFT calculations *(40)*. Alternative pathways leading to the formation of the ether isomer and a potential side product resulting from the reaction with an acetonitrile solvent molecule are also included in the scheme. Green numbers are calculated Gibbs energies (kcal/mol) for each intermediate and transition state relative to **Int1**.

Next, a nucleophilic attack of the generated CF_3^- group on the hypervalent iodine occurs through a concerted transition state where also loss of fluorine from the iodine takes place, yielding Togni reagent **3** complexed with KF. These simultaneous steps occur via **TS3** (Fig. 1 and Scheme 3), which is found to be the rate-determining barrier of this process. Very

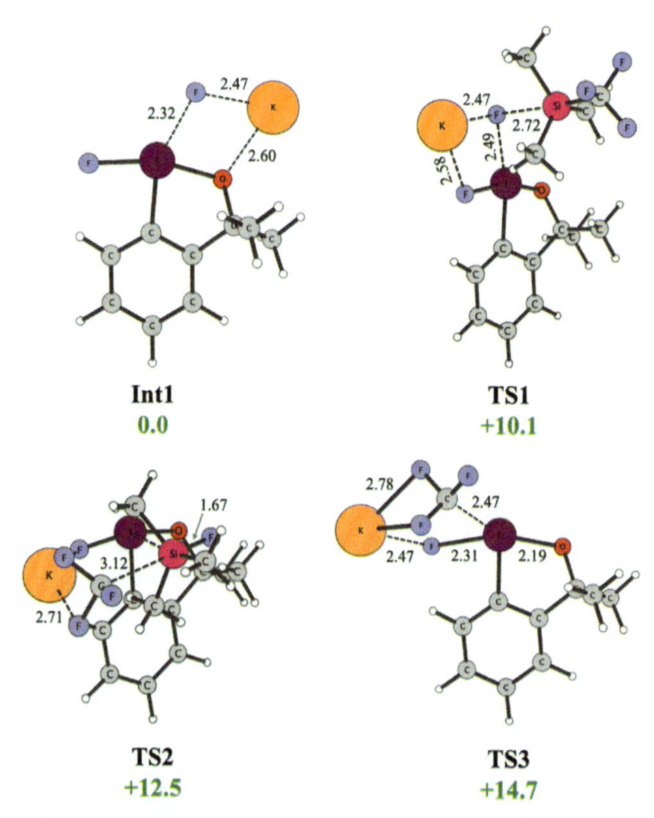

Fig. 1 *Optimized geometries of selected structures involved in the synthesis of Togni reagent* ***3***. Bond distances are given in Å and relative energies are in kcal/mol *(40)*.

interestingly, thus, the calculations show that the KF salt plays a critical role as a catalyst in the reaction, donating a fluorine atom to the Ruppert–Prakash reagent **8** and then accepting a fluorine back from fluoro-benziodoxole reagent **1**.

For the formation of the ether form of the reagent (**Int6**), several possibilities were explored (Scheme 3) *(40)*. Direct isomerization from Togni reagent **3** was found to have very high barriers, even when assisted by KF. Additionally, the potential formation of **Int6** from **Int3** was considered (Scheme 3). However, it was shown that the barrier associated with the second step of this pathway, the electrophilic trifluoromethylation occurring via **TS5**, was higher in energy than **TS3**. Interestingly, from **Int3**, an alternative competitive pathway, which includes the reaction with an acetonitrile solvent molecule, was also found to be feasible (Scheme 3).

Thus, the calculations demonstrated that the mechanism shown in Scheme 3 for the formation of the hypervalent form of the reagent is associated with feasible energy barriers, and, importantly, there is no pathway leading to the ether form, neither from the starting material nor from the Togni reagent **3**. These findings explain why the high-energy hypervalent Togni reagent is kinetically trapped and cannot form the much more stable ether form.

The situation is quite different in the case of the trifluoromethylthio analogue, where a different mechanism was obtained for the synthesis of this reagent, as shown in Scheme 4. The proposed mechanism starts by the formation of a complex between chloro-benziodoxole **7** and $AgSCF_3$ **9** (see **Int9** in Fig. 2 and Scheme 4), where the silver ion facilitates chloride removal from **7**, leading to the generation of an ion-pair intermediate **Int10**. Next, the iodoarene cation undergoes a nucleophilic attack where the SCF_3 group is transferred to the iodine via **TS8**. This process allows the formation of **Int11**, the hypervalent form of the trifluoromethylthio reagent where the

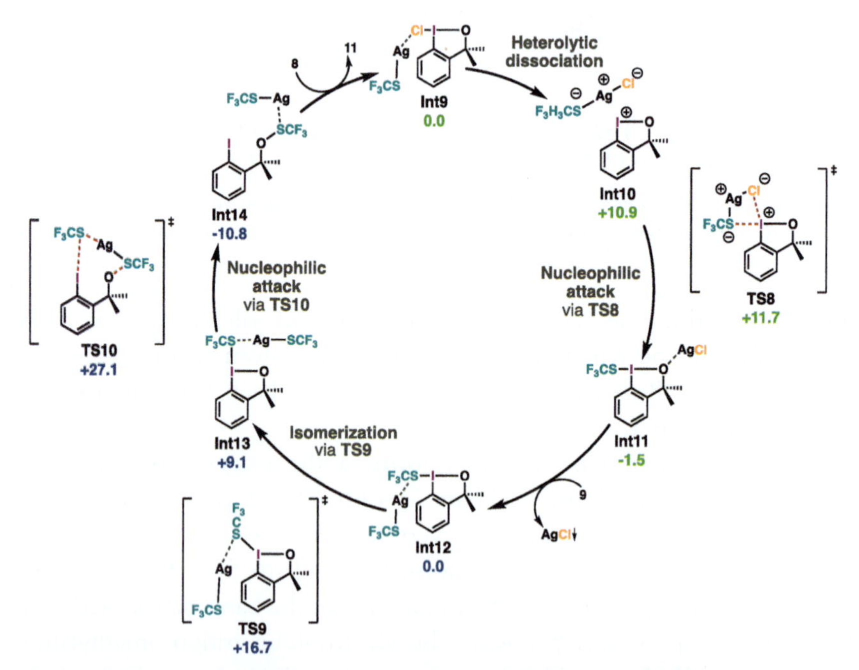

Scheme 4 Proposed reaction mechanism for the synthesis of **11** on the basis of DFT calculations. Green numbers are calculated Gibbs energies (kcal/mol) for intermediates and transition states relative to **Int9**, and blue numbers are calculated Gibbs energies (kcal/mol) for intermediates and transition states relative to **Int12** *(40)*.

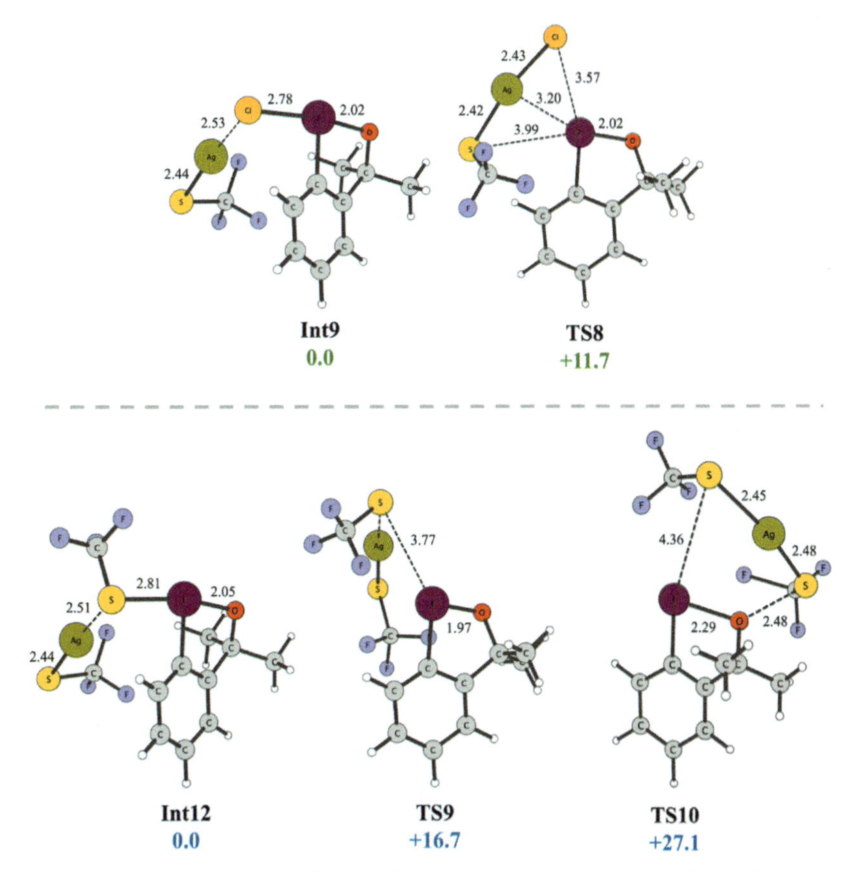

Fig. 2 *Optimized geometries of selected structures involved in the synthesis of **11**.* Bond distances are given in Å, and relative energies are in kcal/mol *(40).*

AgCl is complexed. The overall process of **Int9→Int11** is only slightly exergonic (Scheme 4). However, it is noted that AgCl will precipitate either fully or partially at this juncture. The energetics of this precipitation cannot be evaluated accurately with the employed computational approach, but the overall process is expected to be exothermic and irreversible, and the next step, starting at **Int12**, can thus start at zero energy.

Similar to the synthesis of Togni reagent **3**, the calculations showed that direct isomerization from hypervalent form **10** to the thioperoxide form **11**, or the direct formation of **11** by transfer of the SCF_3 group to the oxygen at either **Int9** or **Int10** are associated with high barriers. However, a major difference compared to the synthesis of the Togni reagent is that after the formation of the hypervalent form, another molecule of $AgSCF_3$ can coordinate (**Int12**) and the silver ion can act as a Lewis acid to facilitate the *trans-cis* isomerization of the F_3CS–I–O bond to yield **Int13** (Scheme 4). Then, the lower energy thioperoxide form (**Int14**) is generated through a nucleophilic attack of the SCF_3 group of the $AgSCF_3$ **7** reagent on the oxygen center simultaneously with the dissociation of the I–O and I–SCF_3 bonds. This process has a feasible energy barrier, explaining why the thioperoxide isomer is observed and not the hypervalent one.

The DFT calculations revealed thus that in both cases, the synthesis of **3** and **11**, that is the hypervalent iodine forms of the reagents, are produced. However, while in the case of **3** the isomerization process is unfeasible due to the high barriers associated with the processes leading to the ether form, for **10** the barrier for its isomerization was found to be energetically feasible under the reaction conditions. These results rationalize the observation of the thermodynamically unstable hypervalent form in the case of the Togni reagent, while this is not observed in the case of the trifluoromethylthio analogue.

The mechanism of the synthesis of a related reagent, Zhang's reagent **13**, was also considered in the computational study. Interestingly, the experimental conditions of the synthesis of this reagent closely resemble those used for the synthesis of **11**, but the hypervalent iodine form **13** was obtained (Scheme 5) *(58)*. The DFT calculations showed that in the synthesis of Zhang's reagent, the catalytic power of the silver is insufficient to catalyze the isomerization process, which was associated with high barriers, resulting thus in the formation of the hypervalent iodine form **13** instead of **14**. The high energy barrier of the isomerization process was attributed to resonance effects due to the acetyl substituent of the reagent *(40)*.

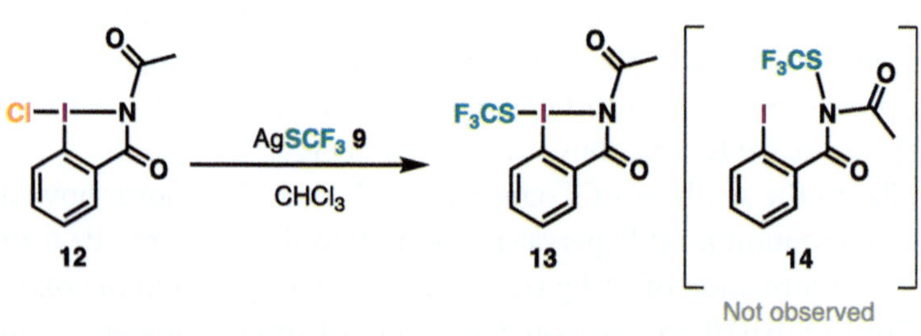

Scheme 5 Synthesis of the fluorine-containing Zhang's reagent **13**.

4. Hypervalent iodine-catalyzed enantioselective fluorocyclization

In this section, we discuss the results of the DFT calculations on the mechanism of the enantioselective iodine(III)-catalyzed electrophilic fluorocyclization of 1,1-disubstituted styrene derivatives developed by Szabó and co-workers (Scheme 6) *(42)*. This reaction yields fluorinated tetrahydrofurans with a tertiary carbon-fluoride stereocenter.

The obtained catalytic cycle is shown in Scheme 7 and begins with the generation of the difluoroiodine(III) species (**Int15**, Fig. 3) through the oxidation and deoxyfluorination of iodoarene **(*R*,*R*)−16**. Although this step was not explicitly addressed in the calculations, the formation of **Int15** is corroborated by experiments confirming the oxidation of iodoarene by *m*CPBA (*meta*-chloroperoxybenzoic acid) to produce the iodosylarene (ArI=O), which is easily transformed into **Int15** upon reaction with py·9HF (hydrogen fluoride pyridine) as a fluorine source *(33–36)*.

Next, the cycle continues with the activation of the difluoroiodine(III) compound **Int15** by the HF source (Scheme 7). Two molecules of HF coordinate to **Int15** to form **Int16**, which after the loss of fluoride via **TS11** yields the cationic fluoroiodonium active catalytic species **Int17**. In **TS11**, one of the carbonyl groups of the side chain of the catalyst facilitates the loss of fluorine through an $I^{+}\cdots O$ interaction (Fig. 3), which also stabilizes the cationic species **Int17** *(59)*. In the calculations, all cationic species were modeled both as ion-pairs with an $(HF)_2F^-$ counterion, or as separate ions. While these two approaches resulted in some significant changes in the energy profiles, they did not impact the mechanistic conclusions. Notably, the model with the ion-pair structures provided the lowest energies *(42)*.

Next, substrate **15** coordinates to the catalyst, and the coordination can occur with either the *Si* or *Re* face of the double bond. It was shown that the coordination to *Re*-face was energetically favorable, resulting in the

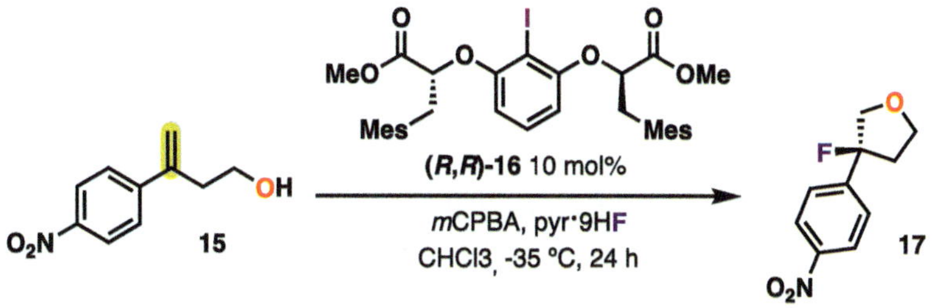

MeO OMe Mes Mes
(*R*,*R*)-16 10 mol%
O_2N OH **15**
*m*CPBA, pyr·9HF
CHCl3, -35 °C, 24 h
F O O_2N **17**

Scheme 6 Hypervalent iodine-catalyzed asymmetric oxyfluorination reaction.

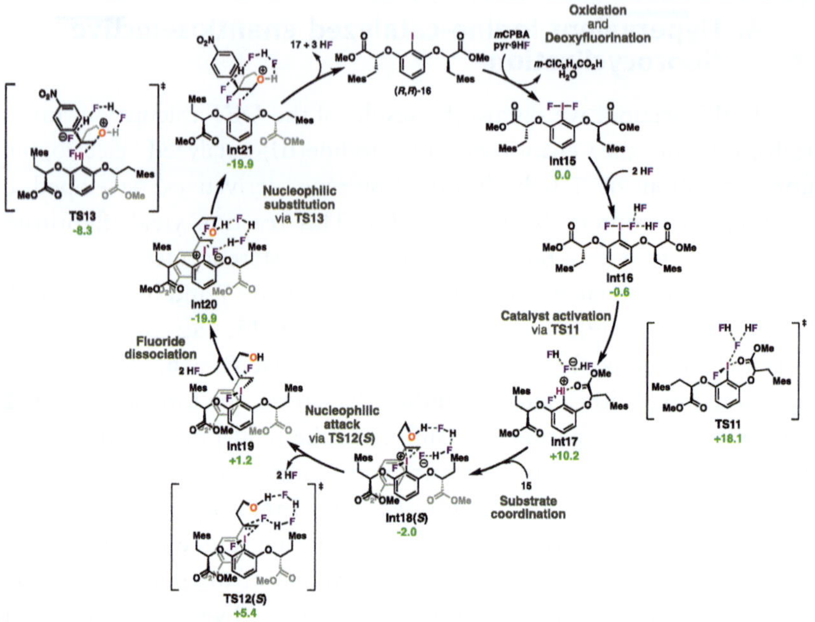

Scheme 7 Catalytic cycle for the hypervalent iodine-catalyzed asymmetric oxyfluorination reaction proposed on the basis of DFT calculations. Green numbers are calculated Gibbs energies (kcal/mol) relative to **Int15** *(42)*.

formation of **Int18(*S*)**, which finally yields the *S*-enantiomer of the product in line with the experimental results. After coordination, a nucleophilic attack by the $(HF)_2F^-$ counterion takes places at the most substituted carbon of the olefin via **TS12(*S*)** to produce **Int19**. Subsequently, **Int19** assisted by two HF molecules undergoes a dissociation of fluoride. This step occurs readily due to the weakening of the I−F bond in **Int19** resulting from the formation of the C-I bond (Scheme 7). The next step is an intramolecular nucleophilic attack by the hydroxyl group of **15**, resulting in the displacement of the aryliodonium moiety and the formation of **Int21**, which contains the protonated form of the cyclic product **17**. The last step of the cycle involves the deprotonation of cyclic compound to yield the final product **17** and regenerate the iodoarene catalyst **(*R*,*R*)−16**.

According to the calculated overall energy profile shown in Fig. 4, the rate-determining step of the catalytic cycle was found to be the activation of the catalyst. The coordination of the substrate (**Int17→Int18(*S*)**) was found to be an irreversible step and constitutes thus the selectivity-determining step of the reaction. However, no transition state could be located for this process, as all optimizations consistently led to **Int18(*S*)**.

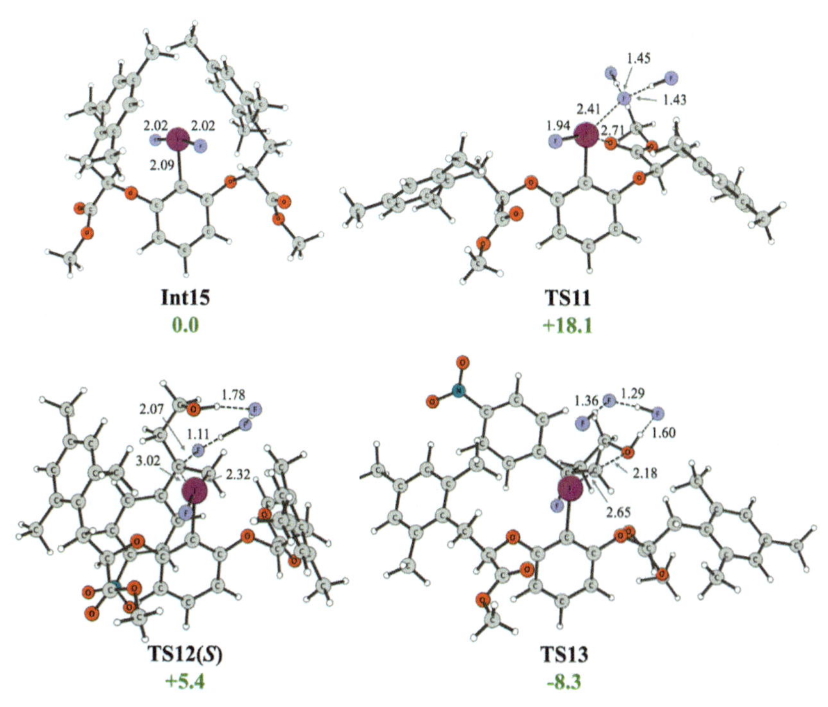

Fig. 3 Optimized geometries of selected structures involved in the hypervalent iodine-catalyzed asymmetric oxyfluorination reaction. Bond distances are given in Å, and relative energies are in kcal/mol *(42)*.

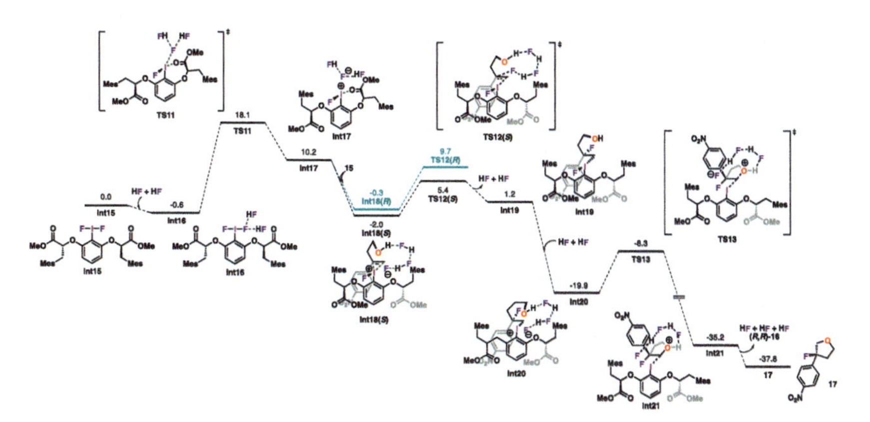

Fig. 4 *Calculated free energy profiles (kcal/*mol*) for the hypervalent iodine-catalyzed asymmetric oxyfluorination reaction (42).*

In order to address this issue and ensure that the correct factors governing the enantioselectivity are identified, the geometry of **Int18(*R*)** was optimized, and was found to be 1.7 kcal/mol higher in energy than **Int18(*S*)**. Then, the potential energy surface was calculated backward from these intermediates by performing constrained optimizations starting from **Int18(*S*)** and **Int18(*R*)**, where the distance between the iodine center and the double bond of the substrate was increased gradually (Fig. 5). This procedure demonstrated that the energy difference between the intermediates is maintained along the reaction coordinate of the coordination, even at long distances resembling the TS structures. This energy difference was almost constant and in good agreement with the experimental outcome (90% ee (*S*)). Thus, the factors governing the enantioselectivity were deduced by analyzing the geometries of **Int18(*S*)** and **Int8(*R*)** (Fig. 5). The analysis revealed that the *ortho*-substituents in the side arm of the catalyst play an important role, since the higher energy associated with **Int18(*R*)** is caused by a steric repulsion between one of these substituents with one of the methylene carbons of substrate **15** *(42)*.

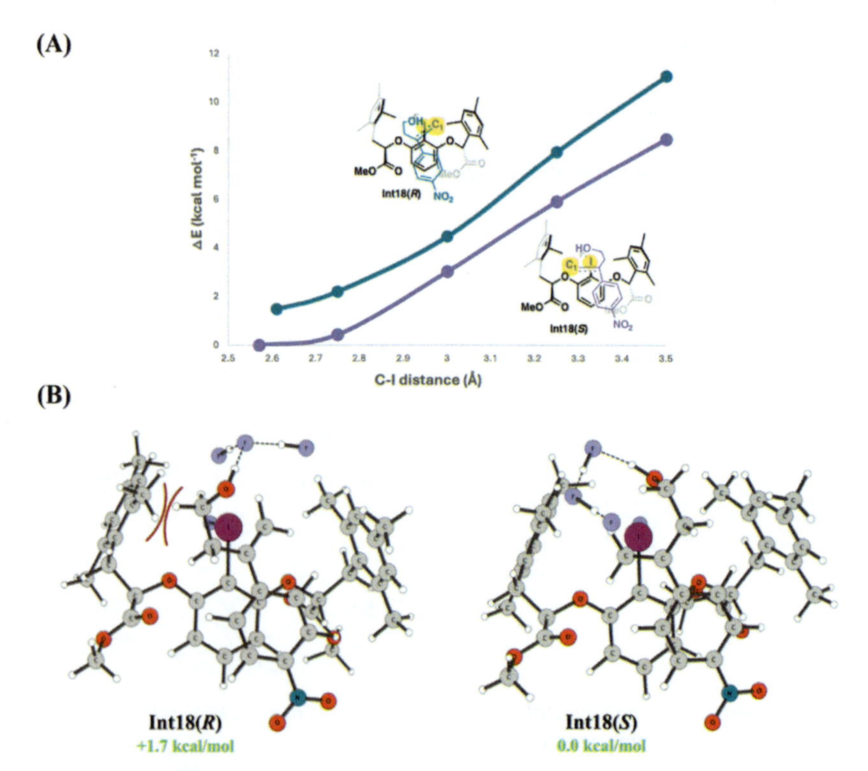

Fig. 5 *(A) Energy as a function of the C_1-I distance, starting from Int18(S) and Int18(R), and (B) optimized geometries of the Int18(S) and Int18(R) (42).*

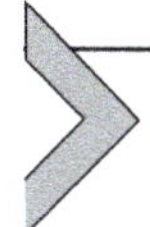

5. Hypervalent iodine-catalyzed bora-wagner-meerwein rearrangement

The last example discussed in this chapter is the electrophilic fluorination of alkenes via bora-Wagner-Meerwein rearrangement catalyzed by in situ generated iodine(III)-catalyst reported by Szabó and co-workers (Scheme 8) *(41)*. This reaction allows the synthesis of β-difluoroalkyl boronates starting from geminal alkyl substituted vinyl-Bmida derivatives.

The catalytic cycle proposed on the basis of the calculations is shown in Scheme 9 *(41)*. As in the previous case (Scheme 7) the first step involves the oxidation and fluorination of iodoarene **19** yielding intermediate **Int22**, and was not explicitly considered by the calculations as it has been corroborated by experimental data *(60)*. In the following steps, the cationic fluoroiodonium intermediate **Int24** is formed through the activation of iodoarene difluoride **Int23**. Similar to the previous case, this activation is modeled by employing two molecules of HF that coordinate to **Int22** to generate **Int23**. This coordination elongates the activated I–F bond, facilitating the subsequent fluoride abstraction that results in the formation of intermediate **Int24** (see Fig. 6).

Once the cationic fluoroiodonium intermediate **Int24** is generated, the coordination of the double bond of Bmida substrate **18** to the iodine atom occurs, resulting in intermediate **Int25** (Scheme 9). Next, the nucleophilic attack of $(HF)_2F^-$ at one of the olefinic carbons of **Int25** takes place. In agreement with the experimentally observed regioselectivity of the reaction, the calculations show that the attack at the most substituted carbon has a barrier of 8 kcal/mol (**TS14**) relative to **Int24**, while the barrier for the attack to the less substituted carbon is 11 kcal/mol.

Also in this case, the formation of the C–I bond at **Int26** facilitates the fluoride dissociation by two molecules of HF to form **Int27**. No transition state could be located for this step. However, considering its exergonicity, the step was estimated to have a very low barrier. Next, the bora-Wagner-Meerwein

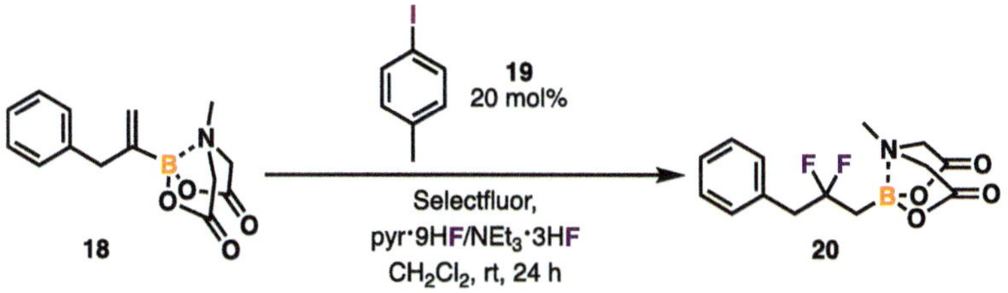

18

19
20 mol%

Selectfluor,
pyr·9HF/NEt$_3$·3HF
CH$_2$Cl$_2$, rt, 24 h

20

Scheme 8 Iodine(III)-catalyzed difluorination via bora-Wagner-Meerwein rearrangement.

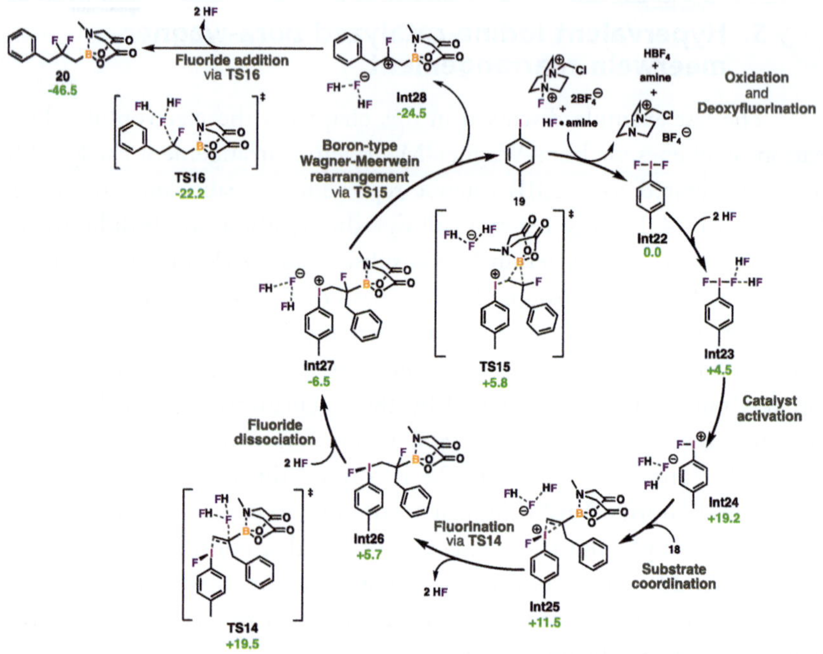

Scheme 9 Proposed catalytic cycle on the basis of DFT calculations for the iodine(III)-catalyzed difluorination via bora-Wagner-Meerwein rearrangement. Green numbers are calculated Gibbs energies (kcal/mol) for intermediates and transition states relative to **Int22** *(41)*.

rearrangement takes place via bora-cyclopropane type structure that resembles those found in the studies performed by the groups of Yudin *(61)* and Pellegrinet *(62)*. This migration step results in the formation of carbocation **Int28** and the regeneration of the iodoarene catalyst **19**.

At this point, the possibility of competing [1,2]-alkyl (benzyl) migration from **Int26** was also considered, but was found to have a barrier 6.9 kcal/mol higher than **TS15**. To complete the catalytic cycle, the **Int27** undergoes a nucleophilic attack by the $(HF)_2F^-$ anion via **TS18** with a very low barrier, yielding the final product **20** *(41)*.

Inspired by the results of Wang and co-workers *(63)* on [1,2]-aryl migration of aryl vinyl-Bmida derivatives, calculations were also performed for representative cases of aryl vs. Bmida group migration. Interestingly, the obtained order of the migration aptitude in electrophilic fluorination of geminal substituted vinyl-Bmida substrates is aryl > Bmida > alkyl, although it was found that the energy differences between the aryl and Bmida migrations are relatively small and also depend on the substituents of the aryl group *(41)*.

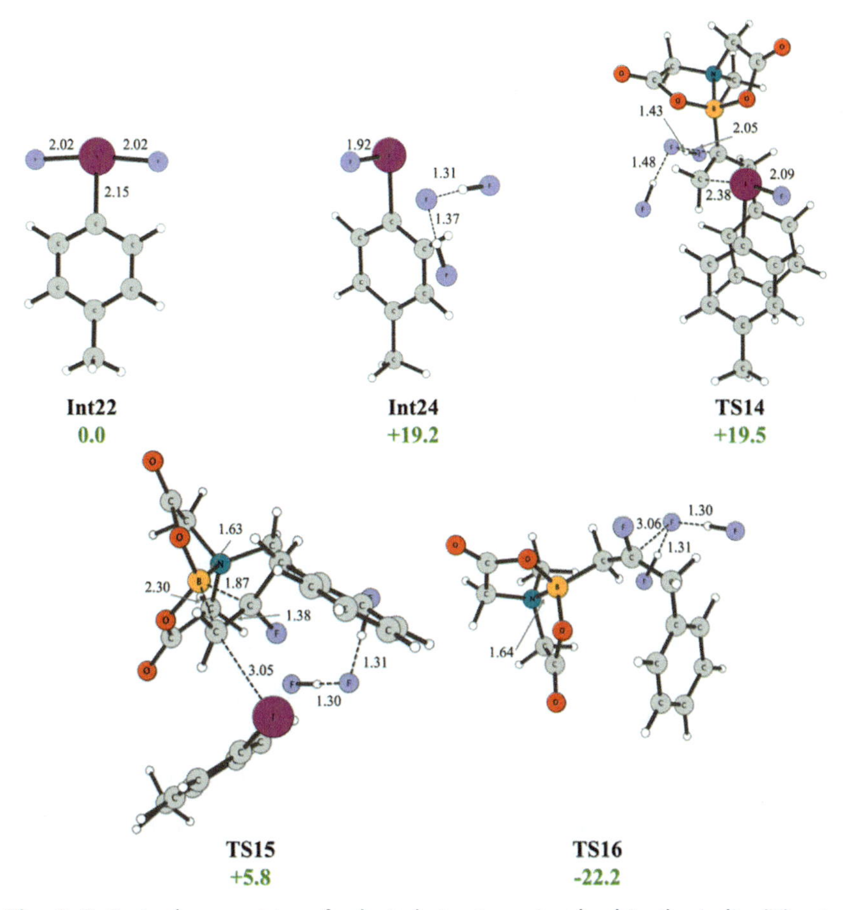

Fig. 6 Optimized geometries of selected structures involved in the iodine(III)-catalyzed difluorination via bora-Wagner-Meerwein rearrangement. Bond distances are given in Å and relative energies are in kcal/mol *(41)*.

6. Conclusions

We have in this chapter discussed a number of representative studies carried out within our research group where DFT calculations have been employed to investigate the reactivity of various fluorine or fluorine-containing transfer hypervalent iodine compounds.

The calculations were able to reproduce and rationalize experimental observations regarding the stability, reactivity, and selectivity of these reagents/catalysts, underscoring the usefulness of the quantum chemical approach in this field. Through detailed investigations and analysis, the calculations offer valuable insights into the properties and reaction mechanisms of these systems that enhance our understanding and pave the way for further innovations in organic synthesis.

Acknowledgments

We thank all co-workers and collaborators who contributed to this work, in particular Dr. Oriana Brea and Prof. Kálmán Szabó. M.B. thanks Ministerio de Ciencia e Innovación for a Juan de la Cierva fellowship. F.H. acknowledges financial support from the Swedish Research Council (2019–04010) and the Knut and Alice Wallenberg Foundation (Dnr: 2018.0066).

References

1. Smart, B. E. Fluorine Substituent Effects (on Bioactivity). *J. Fluor. Chem.* **2001,** *109*, 3–11.
2. O'Hagan, D. Understanding Organofluorine Chemistry. An Introduction to the C–F Bond. *Chem. Soc. Rev.* **2008,** *37*, 308–319.
3. Harsanyia, A.; Sandford, G. Organofluorine Chemistry: Applications, Sources and Sustainability. *Green Chem.* **2015,** *17*, 2081–2086.
4. Müller, K.; Faeh, C.; Diederich, F. Fluorine in Pharmaceuticals: Looking Beyond Intuition. *Science* **2007,** *317*, 1881–1886.
5. Zhou, Y.; Wang, J.; Gu, Z.; Wang, S.; Zhu, W.; Aceña, J. L.; Soloshonok, V. A.; Izawa, K.; Liu, H. Next Generation of Fluorine-Containing Pharmaceuticals, Compounds Currently in Phase II–III Clinical Trials of Major Pharmaceutical Companies: New Structural Trends and Therapeutic Areas. *Chem. Rev.* **2016,** *116*, 422–518.
6. Zhu, Y.; Han, J.; Wang, J.; Shibata, N.; Sodeoka, M.; Soloshonok, V. A.; Coelho, J. A. S.; Toste, F. D. Modern Approaches for Asymmetric Construction of Carbon–fluorine Quaternary Stereogenic Centers: Synthetic Challenges and Pharmaceutical Needs. *Chem. Rev.* **2018,** *118*, 3887–3964.
7. Inoue, M.; Sumii, Y.; Shibata, N. Contribution of Organofluorine Compounds to Pharmaceuticals. *ACS Omega* **2020,** *5*, 10633–10640.
8. He, J.; Li, Z.; Dhawan, G.; Zhang, W.; Sorochinsky, A. E.; Butler, G.; Soloshonok, V. A.; Han, J. Fluorine-containing Drugs Approved by the FDA in 2021. *Chin. Chem. Lett.* **2022,** *34*, 107578.
9. Hana, Z.-Z.; Zhanga, C.-P. Fluorination and Fluoroalkylation Reactions Mediated by Hypervalent Iodine Reagents. *Adv. Synth. Catal.* **2020,** *362*, 4256–4292.
10. Szabó, K. J. Fluorination, Trifluoromethylation, and Trifluoromethylthiolation of Alkenes, Cyclopropanes, and Diazo Compounds. In *Organofluorine Chemistry;* Szabó, K., Selander, N., Eds.; Wiley: Germany, 2021.
11. Charpentier, J.; Früh, N.; Togni, A. Electrophilic Trifluoromethylation by use of Hypervalent Iodine Reagents. *Chem. Rev.* **2015,** *115*, 650–682.
12. Singh, F. V.; Shetgaonkar, S. E.; Krishnan, M.; Wirth, T. Progress in Organocatalysis with Hypervalent Iodine Catalysts. *Chem. Soc. Rev.* **2022,** *51*, 8102–8139.
13. Arnold, A. M.; Ulmer, A.; Gulder, T. Advances in Iodine (III)-Mediated Halogenations: A Versatile Tool to Explore New Reactivities and Selectivities. *Chem. Eur. J.* **2016,** *22*, 8728–8739.
14. Kohlheppa, S. V.; Gulder, T. Hypervalent Iodine (III) Fluorinations of Alkenes and Diazo Compounds: New Opportunities in Fluorination Chemistry. *Chem. Soc. Rev.* **2016,** *45*, 6270–6288.
15. Musher, J. I. The Chemistry of Hypervalent Molecules. *Angew. Chem., Int. Ed. Engl.* **1969,** *8*, 54–68.

16. Merritt, E. A.; Olofsson, B. Diaryliodonium Salts: A Journey from Obscurity to Fame. *Angew. Chem. Int. Ed.* **2009,** *48*, 9052–9070.
17. Yoshimura, A.; Zhdankin, V. V. Advances in Synthetic Applications of Hypervalent Iodine Compounds. *Chem. Rev.* **2016,** *116*, 3328–3435.
18. Uyanik, M.; Ishihara, K. Hypervalent Iodine Catalysis by Organocatalysts. *Chem. Commun.* **2013,** *49*, 2084–2097.
19. Le, Du, E.; Waser, J. Recent Progress in Alkynylation with Hypervalent Iodine Reagents. *Chem. Commun.* **2023,** *59*, 1589–1604.
20. Yoshimura, A.; Zhdankin, V. V. Advances in Synthetic Applications of Hypervalent Iodine Compounds. *Chem. Rev.* **2016,** *116*, 3328–3435.
21. Li, Y.; Hari, D. P.; Vita, M. V.; Waser, J. Cyclic Hypervalent Iodine Reagents for Atom-Transfer Reactions: Beyond Trifluoromethylation. *Angew. Chem. Int. Ed.* **2016,** *55*, 4436–4454.
22. Chen, J.-Y.; Huang, J.; Sun, K.; He, W.-M. Recent Advances in Transition-metal-free Trifluoromethylation with Togni's Reagents. *Org. Chem. Front.* **2022,** *9*, 1152–1164.
23. Yoshimura, A.; Saito, A.; Zhdankinc, V. V. Recent Progress in Synthetic Applications of Cyclic Hypervalent Iodine(III) Reagents. *Adv. Synth. Catal.* **2023,** *365*, 2653–2675.
24. Andries-Ulmer, A.; Brunner, C.; Rehbein, J.; Gulder, T. Fluorine as a Traceless Directing Group for the Regiodivergent Synthesis of Indoles and Tryptophans. *J. Am. Chem. Soc.* **2018,** *140*, 13034–13041.
25. Zhou, B.; Yan, T.; Xue, X.-S.; Cheng, J.-P. Mechanism of Silver-Mediated Geminal Difluorination of Styrenes with a Fluoroiodane Reagent: Insights into Lewis-Acid-Activation Model. *Org. Lett.* **2016,** *18*, 6128–6131.
26. Zhou, B.; Xue, X.-S.; Cheng, J.-P. Theoretical Study of Lewis Acid Activation Models for Hypervalent Fluoroiodane Reagent: The Generality of "F-coordination" Activation Model. *Tetrahedron Lett.* **2017,** *58*, 1287–1291.
27. Zhang, J.; Szabó, K. J.; Himo, F. Metathesis Mechanism of Zinc-catalyzed Fluorination of Alkenes with Hypervalent Fluoroiodine. *ACS Catal.* **2017,** 7, 1093–1100.
28. Ling, L.; Liu, K.; Li, X.; Li, Y. General Reaction Mode of Hypervalent Iodine Trifluoromethylation Reagent: A Density Functional Theory Study. *ACS Catal.* **2015,** *5*, 2458–2468.
29. Mai, B. K.; Kálmán, J.; Szabó, K. J.; Himo, F. Mechanisms of Rh-catalyzed Oxyfluorination and Oxytrifluoromethylation of Diazocarbonyl Compounds with Hypervalent Fluoroiodine.*ACS Catal.* **2018,** *8*, 4483–4492.
30. Pinto de Magalhães, H.; Togni, A.; Lüthi, H. P. *J. Org. Chem.* **2017,** *82*, 11799–11805.
31. Sala, O.; Santschi, N.; Jungen, S.; Lüthi, H. P.; Iannuzzi, M.; Hauser, N.; Togni, A. S-Trifluoromethylation of Thiols by Hypervalent Iodine Reagents: A Joint Experimental and Computational Study. *Chem. Eur. J.* **2016,** *22*, 1704–1713.
32. Yang, W.; Ma, D.; Zhou, Y.; Dong, X.; Lin, Z.; Sun, J. NHC-Catalyzed Electrophilic Trifluoromethylation: Efficient Synthesis of γ-Trifluoromethyl α,β-Unsaturated Esters. *Angew. Chem. Int. Ed.* **2018,** *57* 12097–1210.
33. Suzuki, S.; Kamo, T.; Fukushi, K.; Hiramatsu, T.; Tokunaga, E.; Dohi, T.; Kita, Y.; Shibata, N. Iodoarene-catalyzed Fluorination and Aminofluorination by an Ar-I/HF·Pyridine/mCPBA System. *Chem. Sci.* **2014,** *5*, 2754–2760.
34. Arrica, M. A.; Wirth, T. Fluorinations of α-Seleno Carboxylic Acid Derivatives with Hypervalent (Difluoroiodo) toluene. *Eur. J. Org. Chem.* **2005,** *2005*, 395–403.
35. Kitamura, T.; Kuriki, S.; Morshed, M. H.; Hori, Y. A Practical and Convenient Fluorination of 1, 3-dicarbonyl Compounds Using Aqueous hf in the Presence of Iodosylbenzene. *Org. Lett.* **2011,** *13*, 2392–2394.
36. Kitamura, T.; Muta, K.; Kuriki, S. Catalytic Fluorination of 1, 3-dicarbonyl Compounds Using Iodoarene Catalysts. *Tetrahedron Lett.* **2013,** *54*, 6118–6120.

37. Banik, S. M.; Medley, J. W.; Jacobsen, E. N. Catalytic, Asymmetric Difluorination of Alkenes to Generate Difluoromethylated Stereocenters. *Science*. **2016,** *353*, 51–54.
38. Xu, P.; Wang, F.; Fan, G.; Xu, X.; Tang, P. Hypervalent Iodine(III)-Mediated Oxidative Fluorination of Alkylsilanes by Fluoride Ions. *Angew. Chem.* **2017,** *129*, 1121–1124.
39. Su, J.; Shu, S.; Li, Y.; Chen, Y.; Dong, J.; Liu, Y.; Fang, Y.; Ke, Z. Mechanism-Dependent Selectivity: Fluorocyclization of Unsaturated Carboxylic Acids or Alcohols by Hypervalent Iodine. *Front. Chem.* **2022,** *10*, 897828.
40. Brea, O.; Szabo, K. J.; Himo, F. Mechanisms of Formation and Rearrangement of Benziodoxole-Based CF_3 and SCF_3 Transfer Reagents.*J. Org. Chem.* **2020,** *85*, 15577–15585.
41. Wang, Q.; Biosca, M.; Himo, F.; Szabó, K. J. Electrophilic Fluorination of Alkenes via Bora-Wagner–Meerwein Rearrangement. Access to β-Difluoroalkyl Boronates. *Angew. Chem. Int. Ed.* **2021,** *60*, 26327–26331.
42. Wang, Q.; Lübcke, M.; Biosca, M.; Hedberg, M.; Eriksson, L.; Himo, F.; Szabó, K. J. Enantioselective Construction of Tertiary Fluoride Stereocenters by Organocatalytic Fluorocyclization. . *J. Am. Chem. Soc.* **2020,** *142*, 20048–20057.
43. Becke, A. D. Density-functional Exchange-energy Approximation with Correct Asymptotic Behavior. *Phys. Rev. A* **1988,** *38*, No. 3098.
44. Becke, A. D. Density-Functional Thermochemistry. III. The Role of Exact Exchange. *J. Chem. Phys.* **1993,** *98*, 5648–5652.
45. Grimme, S.; Antony, J.; Ehrlich, S.; Krieg, H. A Consistent and Accurate Ab Initio Parametrization of Density Functional Dispersion Correction (DFT-D) for the 94 Elements H-Pu. *J. Chem. Phys.* **2010,** *132* No. 154104.
46. Grimme, S.; Ehrlich, S.; Goerigk, L. Effect of the Damping Function in Dispersion Corrected Density Functional Theory. *J. Comput. Chem.* **2011,** *32*, 1456–1465.
47. Marenich, A. V.; Cramer, C. J.; Truhlar, D. G. Universal Solvation Model Based on Solute Electron Density and on a Continuum Model of the Solvent Defined by the Bulk Dielectric Constant and Atomic Surface Tensions. *J. Phys. Chem. B* **2009,** *113*, 6378–6396.
48. Besora, M.; Braga, A. A. C.; Ujaque, G.; Maseras, F.; Lledos, A. The Importance of Conformational Search: A Test Case on the Catalytic Cycle of the Suzuki-Miyaura Cross-coupling. *Theor. Chem. Acc.* **2011,** *128*, 639–646.
49. Friedrich, N.-O.; De Bruyn Kops, C.; Flachsenberg, F.; Sommer, K.; Rarey, M.; Kirchmair, J. Benchmarking Commercial Conformer Ensemble Generators. *J. Chem. Inf. Model.* **2017,** *57*, 2719–2728.
50. Brethomé, A. V.; Fletcher, S. P.; Paton, R. S. Conformational Effects on Physical-Organic Descriptors: The Case of Sterimol Steric Parameters. *ACS Catal.* **2019,** *9*, 2313–2323.
51. Balcells, D.; Drudis-Sole, G.; Besora, M.; Dolker, N.; Ujaque, G.; Maseras, F.; Lledos, A. Some Critical Issues in the Application of Quantum Mechanics/molecular Mechanics Methods to the Study of Transition Metal Complexes. *Faraday Discuss* **2003,** *124*, 429–441.
52. Bartol, J.; Comba, P.; Melter, M.; Zimmer, M. Conformational Searching of Transition Metal Compounds. *J. Comput. Chem.* **1999,** *20*, 1549–1558.
53. Matoušěk, V.; Pietrasiak, E.; Schwenk, R.; Togni, A. One-pot Synthesis of Hypervalent Iodine Reagents for Electrophilic Trifluoromethylation. *J. Org. Chem.* **2013,** *78*, 6763–6768.
54. Shao, X.; Wang, X.; Yang, T.; Lu, L.; Shen, Q. An Electrophilic Hypervalent Iodine Reagent for Trifluoromethylthiolation. *Angew. Chem., Int. Ed.* **2013,** *52*, 3457–3460.
55. Vinogradova, E. V.; Müller, P.; Buchwald, S. L. Structural Reevaluation of the Electrophilic Hypervalent Iodine Reagent for Trifluoromethylthiolation Supported by the Crystalline Sponge Method for X-ray Analysis. *Angew. Chem., Int. Ed.* **2014,** *53*, 3125–3128.

56. Sun, T.-Y.; Wang, X.; Geng, H.; Xie, Y.; Wu, Y.-D.; Zhang, X.; Schaefer, H. F. Why does Togni's Reagent I Exist in the High-energy Hypervalent Iodine Form? Re-evaluation of Benziodoxole Based Hypervalent Iodine Reagents. *Chem. Commun.* **2016,** *52*, 5371–5374.
57. Koichi, S.; Leuthold, B.; Lüthi, H. P. Why do the Togni Reagent and Some of its Derivatives Exist in the High-energy Hypervalent Iodine Form? New Insight into the Origins of their Kinetic Stability. *Phys. Chem.* **2017,** *19*, 32179–32183.
58. Yang, X.-G.; Zheng, K.; Zhang, C. Electrophilic Hypervalent Trifluoromethylthio-iodine (III) Reagent. *Org. Lett.* **2020,** *22*, 2026–2031.
59. (The same mechanism was found for the aryl iodine-catalyzed asymmetric difluorination of β-substituted styrenes) Zhou, B.; Haj, M. K.; Jacobsen, E. N.; Houk, K. N.; Xue, X.-S. Mechanism and Origins of Chemo-and Stereoselectivities of Aryl Iodide-catalyzed Asymmetric Difluorinations of β-substituted Styrenes. *J. Am. Chem. Soc.* **2018,** *140*, 15206–15218.
60. Sarie, J.; Thiehoff, C.; Mudd, R.; Daniliuc, C.; Kehr, G.; Gilmour, R. Deconstructing the Catalytic, Vicinal Difluorination of Alkenes: HF-Free Synthesis and Structural Study of p-TolIF2. *J. Org. Chem.* **2017,** *82*, 11792–11798.
61. Lee, C. F.; Diaz, D. B.; Holownia, A.; Kaldas, S. J.; Liew, S. K.; Garrett, G. E.; Dudding, T.; Yudin, A. K. Amine Hemilability Enables Boron to Mechanistically Resemble either Hydride or Proton. *Nat. Chem.* **2018,** *10*, 1062–1070.
62. Vallejos, M. M.; Pellegrinet, S. C. Theoretical Study of the BF3-Promoted Rearrangement of Oxiranyl N-Methyliminodiacetic Acid Boronates. *J. Org. Chem.* **2017,** *82*, 5917–5925.
63. Lv, W.-X.; Li, Q.; Li, J.-L.; Li, Z.; Lin, E.; Tan, D.-H.; Cai, Y.-H.; Fan, W.-X.; Wang, H. gem-Difluorination of Alkenyl N-methyliminodiacetyl Boronates: Synthesis of α- and β-Difluorinated Alkylborons. *Angew. Chem. Int. Ed.* **2018,** *57*, 16544–16548.

About the authors

Maria Biosca received her Ph.D. in 2018 at University Rovira i Virgili (URV) under the supervision of Profs. M.Diéguez and O. Pàmies. During her Ph.D., she did a short exchange in the group of Prof. M. Alcarazo (Göttingen University). In 2019, she joined Prof. F. Himo and Prof. K. J. Szabó's groups at Stockholm University as a postdoctoral researcher. In 2022, she came back to URV as a Juan de la Cierva postdoctoral fellow, to work in the groups of Profs. M. Diéguez and J. M. Poblet.

Fahmi Himo obtained his Ph.D. at Stockholm University, Sweden, in 2000. He then received a postdoctoral fellowship from the Wenner-Gren Foundations for spending two years at The Scripps Research Institute in La Jolla, USA, and three years at the Royal Institute of Technology (KTH) in Stockholm. He became a professor at Stockholm University in 2009.

CHAPTER TWO

Combining DFT and experimental studies in enantioselective catalysis: From rationalization to prediction

Maria Biosca[a,*], Maria Besora[a], Feliu Maseras[b], Oscar Pàmies[a], and Montserrat Diéguez[a,*]

[a]Departament de Química Física i Inorgànica, Universitat Rovira i Virgili, Tarragona, Spain
[b]Institute of Chemical Research of Catalonia (ICIQ), The Barcelona Institute of Science and Technology, Tarragona, Spain
*Corresponding authors. e-mail address: maria.biosca@urv.cat; montserrat.dieguez@urv.cat

Contents

Abstract

Asymmetric catalysis is key for the achievement of important chemical products with application in the pharmaceutical and agrochemical industry. Among the various types of asymmetric catalyzed reactions, metal-catalyzed reactions are widely employed, with a diverse range of chiral catalysts available nowadays. Chiral catalysts play a crucial role in chirality induction during the formation of new bonds. Despite this diversity, certain reactions still need the development of new catalysts to enhance reactivities and selectivities of specific substrates. Nevertheless, the pursuit of these optimal catalytic systems often presents major challenges, demanding significant time and resources.

The key of enantioselectivity control lies in the lowest energy transition states of the enantioselective-determining step, and will dictate the stereochemical outcome of the reaction. Although repulsive steric factors have traditionally accounted for the

Advances in Catalysis, Volume 75
ISSN 0360-0564, https://doi.org/10.1016/bs.acat.2024.08.002

stability of these transition states, modern understanding includes the role of attractive forces, such as weak non-covalent interactions. Current computational methods, such as density functional theory calculations (DFT) can be applied to large systems, enabling the analysis of both factors, enhancing our comprehension of stereoselectivity origins.

In this chapter, we present several illustrative examples from our recent research, using DFT calculations to elucidate the underlying factors behind enantioselectivity. Our focus was on two specific asymmetric metal-catalyzed reactions: Ir-hydrogenation of non-chelating olefins and Pd-allylic substitution. Our investigations provided new insights into the factors governing the enantioselectivity of these reactions with various chiral catalysts including P,N- and P,S-based ligands. In some cases, our findings have allowed for the prediction of catalyst enantioselectivities before their synthesis, thereby accelerating the catalyst design process.

1. Introduction

The ever-increasing demand of chiral molecules, which are present in the synthetic core of many industrial products, has stimulated the interest of chemical community for the search of new or improved synthetic routes. Homogeneous asymmetric metal-catalyzed reactions stands out as a highly effective and selective method for synthesizing these compounds *(1–4)*. In order to obtain excellent results in the enantioselective reactions, several parameters need to be taken into account, although the design and development of the chiral ligand is perhaps one of the most crucial. Therefore, the synthesis of new chiral ligands is essential for the development of catalytic reactions that provide the desired chiral compounds in high yields and enantioselectivities. However, the quest for these optimal catalytic systems often proves to be challenging, time-intensive and costly. Hence, the search of new methodologies that aid in this ligand design is highly advantageous *(1–4)*.

Thanks to the remarkable advancements in the last times in the field of computational chemistry, quantum chemical tools have achieved a significant role in the study of asymmetric catalytic reactions *(5–8)*. Density Functional Theory (DFT) calculations, in particular, have seen widespread use for this purpose. Its use is not only limited to elucidate complex reaction mechanisms, that sometimes can be elusive through experimental methods, but can also be employed to understand the factors governing the enantioselectivity, enabling a further rational design of the catalytic system. Furthermore, they can go even a step forward predicting catalyst reactivities and enantioselectivities, thereby allowing the improvement in terms of speed, cost, and environmental impact of the catalyst design process.

In this context, this book chapter will summarize our recent efforts in the application of DFT calculations in the Ir-catalyzed asymmetric hydrogenation of non-chelating olefins *(9–12)* and the Pd-catalyzed allylic substitution reactions *(13–15)* to assist the design of the chiral catalysts (Scheme 1). We will discuss a number of examples, ranging from more simple rationalizations of enantioselectivity to more intricate scenarios involving catalyst prediction. Prior to delving into the main content of this book chapter, we will offer a brief summary of the most relevant mechanistic aspects of both enantioselective reactions. As it is crucial to identify the step that dictates the enantioselectivity in both mechanisms, as it is fundamental for the subsequent discussion.

In summary, the objective of this book chapter is to provide a closer and simple insight into the usefulness of computational tools in the field of homogeneous asymmetric metal-catalyzed reactions. Furthermore, we aim to demonstrate how the development and design of catalytic systems can be enhanced by the combination of computational and experimental approaches.

2. Technical details

All results discussed in this book chapter were achieved using B3LYP functional *(16–18)*. This methodology has seen extensive application in the exploration of numerous organometallic reactions. Concerning the basis sets employed for geometry optimization, effective core potentials *(19,20)* were utilized for metals, while 6–31G* *(21–23)* or 6–31G** *(24)* were employed for other atoms. Implicit solvation was included into geometry optimizations using the PCM method with the parameters for dichloromethane *(25)*. For enhanced accuracy, single-point calculations conducted based on the optimized structures, using the same basis set for metals

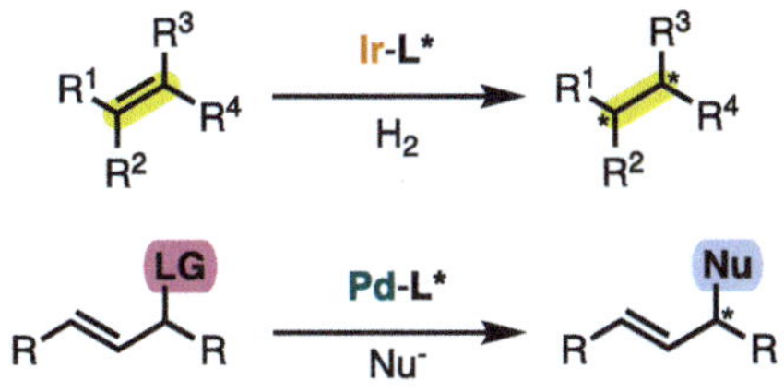

Scheme 1 Metal-catalyzed reactions considered in the current book chapter: Ir-catalyzed asymmetric hydrogenation of non-chelating olefins (1) and Pd-catalyzed asymmetric allylic substitution reaction (2).

and the 6–311+G** *(26,27)* basis set for all other elements. All reported energies are Gibbs free energies in solution.

A crucial point to highlight here is that when trying to reproduce or predict the enantioselectivity of a reaction, small errors in the energy difference between the two most stable transition states, leading opposite enantiomers, can result into notable shifts in the predicted enantioselectivity value. For most catalytic systems due to the similar nature of the transition states leading to both enantiomers, error cancellation helps to achieve good computational results *(28)*. This is because we are not trying to predict the overall barrier weights, but just barrier differences for similar transition. To have a good computational estimation of the *ee*, it's crucial to consider not only the two lowest transition states, but all transition states potentially involved in the enantio-determining step, along with their respective isomers. Additionally, a meticulous conformational analysis is essential for accurately calculating all arrangements of these transition states.

To rationalize the factors governing the enantioselectivity, several on purpose tools have been developed, such as SamVuca *(29)* or NEST*(30)*, but we have commonly used two programs; MolQuo software *(31,32)* (for analysis of steric factors) and NCI plot *(33)* (for non-covalent interactions analysis). MolQuO program allows the creation of quadrant model representations, offering numerical insight into the occupancy of each quadrant. It is noteworthy that the analysis is typically conducted by taking the geometry of the whole TS, but considering only the half of the catalyst where the reactivity takes place and removing the atoms of the olefin in the MolQuO calculation. In contrast, the NCI-plot, is used to map regions in real-space where non-covalent interactions play a significant role. This approach relies solely on electron density and its derivatives, and identifies non-covalent interactions from the reduced density gradient at low densities. It's important to note that the information derived from NCI plots is primarily qualitative.

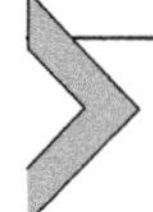

3. Ir-catalyzed asymmetric hydrogenation of non-chelating olefins

Asymmetric metal-catalyzed hydrogenation of olefins is one of the most powerful tools for producing enantiomerically pure compounds *(1–4,34–36)*. While the hydrogenation of olefins containing an adjacent polar group by Rh(I) and Ru(II) complexes bearing phosphorus ligands boasts a long-established history *(1–4,37–39)*, the reduction of non-chelating olefins remains relatively underdeveloped. The high stereoselectivity achieved in the asymmetric hydrogenation of alkenes using Rh/Ru-phosphorus ligands rely on the presence of a coordinating group in a close proximity to the double bond. This groups serves as a secondary complexation function, supplementing the alkene functionality to form a chelate that restricts available conformations *(40)*. In contrast, iridium complexes featuring chiral heterodonor P,X-ligands (where $X = N$, S or O) have emerged as efficient catalysts for asymmetric hydrogenation of non-chelating olefins. Notably, iridium catalysts do not require the presence of an additional coordinating group, rendering their scope complementary to that of Rh- and Ru-complexes with phosphorus ligands *(41–50)*.

3.1 Overview of mechanistic aspects

The reaction mechanisms of the Ir-catalyzed asymmetric hydrogenation of non-chelating olefins have been explored both experimentally and computationally. Initially, four different mechanisms have been proposed; two of them involve Ir(I)/Ir(III) intermediates *(51,52)* and the other two Ir (III)/Ir(V) species *(53–55)*.

Andersson and co-workers employed DFT calculations along with a comprehensive set of experimentally tested ligands and substrates to study all potential diastereomeric pathways of the four proposed mechanisms *(56)*. This extensive study revealed that the catalytic cycle proceeded through Ir(III)/Ir(V) intermediates. However, they encountered difficulties in distinguishing between the two Ir(III)/Ir(V) mechanisms. One mechanism involves migratory insertion of a hydride onto the olefin followed by reductive elimination (Schemes 2, **3/5-MI**), while the other mechanism entailed σ-bond metathesis followed by reductive elimination (Schemes 2, **3/5-Meta**). Noteworthy, it was demonstrated that the transition states (TSs) responsible for the enantiocontrol of the reaction are the migratory insertion transition state (TS) in **3/5-MI** (**TS**$_{\mathbf{MI}}$) and σ-bond metathesis TS in **3/5-Meta** (**TS**$_{\mathbf{Meta}}$). Later on, other research groups reported further computational investigations using iridium catalysts with different P,N-ligands *(57,58)*. These studies provided additional confirmation that the hydrogenation of non-chelating olefins predominantly follows the **3/5-MI** pathway. Nonetheless, the energetic barriers between both catalytic cycles were found to be similar, making it difficult to directly discard the **3/5-Meta** pathway.

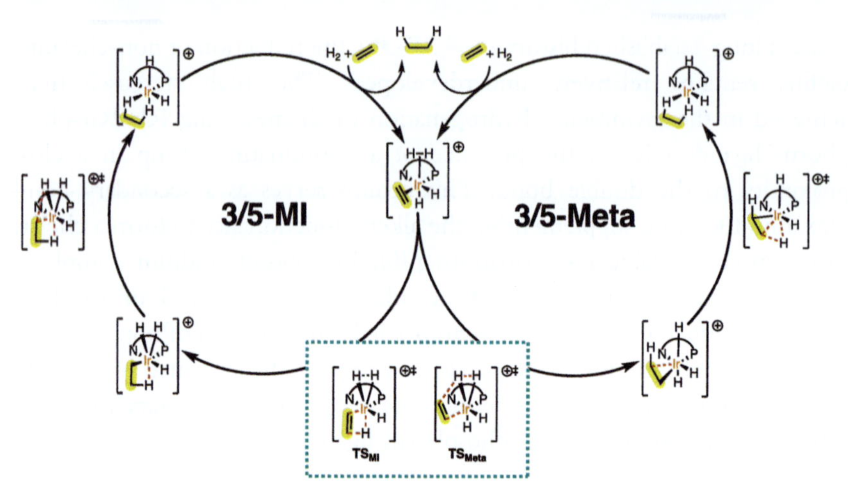

Scheme 2 Proposed catalytic cycles **3/5-MI** and **3/5-Meta** for asymmetric hydrogenation of non-chelating olefins.

More recently, Pfaltz and co-workers conducted an experimental study employing NMR techniques that enabled the characterization of the elusive catalytically active intermediates of the reaction *(59)*. The study revealed that the resting state of the reaction is Ir-dihydride alkene complex ($[Ir(H)_2(alkene)(L)]^+$). Additionally, it was observed that an additional molecule of H_2 was necessary to convert the catalyst-bound alkene into the hydrogenated product. Both findings further reinforce the Ir(III)/Ir(V) mechanism via an $[Ir(H)_2(alkene)(H_2)(L)]^+$ intermediate.

Based on these precedents, to computationally assess the enantioselectivity of the reaction, it is essential to compute all isomers of the $\mathbf{TS_{MI}}$ and $\mathbf{TS_{Meta}}$ (Fig. 1 and Scheme 2). This entails considering the relative rearrangement of the hydride (up or down), the coordination of the olefin through the *Si*-face or *Re*-face and the attack of the hydride through the two olefinic carbons (C_1 and C_2). Additionally, a conformational exploration of all the possible conformation of the chelating ring of all these key isomers is necessary for each individual case.

3.2 Representative examples

The first example discussed here about the origin of the enantioselectivity on the asymmetric reduction of non-chelating olefins is associated with a family of Ir-catalysts containing thioether-phosphite/phosphinite ligands *(10)*. Preliminary experimental results in the asymmetric hydrogenation of the model trisubstituted olefin **S1** indicates that the enantioselectivity is

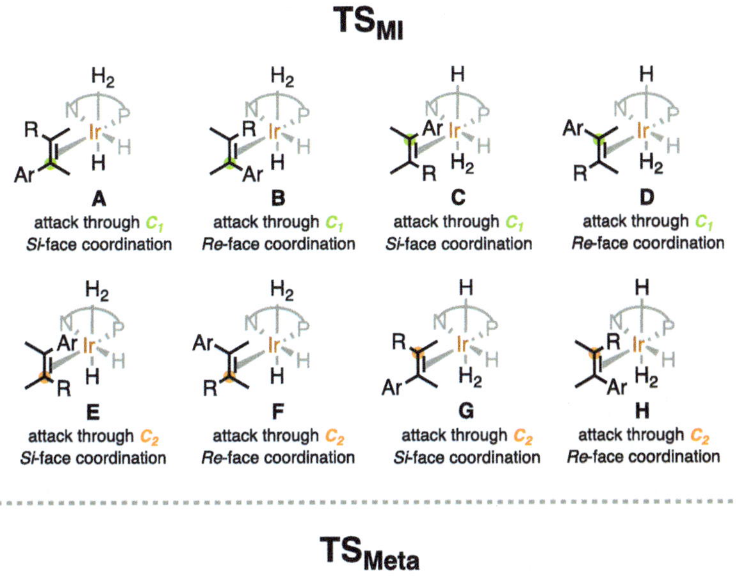

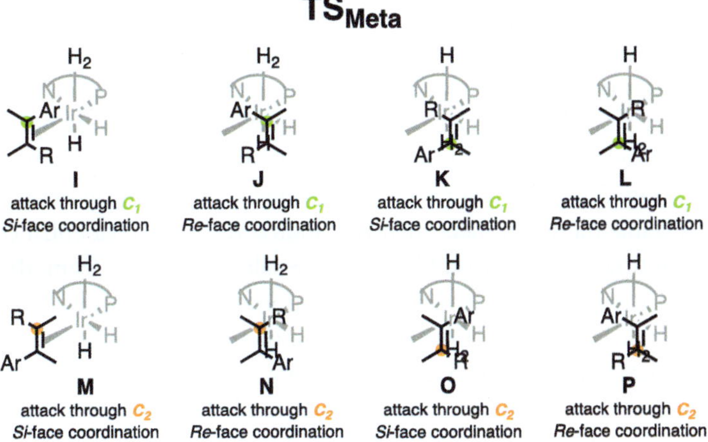

Fig. 1 Key isomers of $\mathbf{TS_{MI}}$ and $\mathbf{TS_{Meta}}$ of Ir-catalyzed asymmetric hydrogenation of non-chelating olefins.

predominantly affected by the type of P-donor group and the thioether substituent (Scheme 3). The best results were obtained with a ligand containing a bulky di-*o*-tolyl phosphinite moiety and a 2,6-dimethylphenyl thioether substituent.

In order to gain a deeper understanding on the factors influencing the stereochemical outcome and to facilitate further rationalization of the catalyst design, a DFT study was performed. Ligands **L1** and **L2** were selected because they contain the simple, unsubstituted diphenyl phosphite moiety,

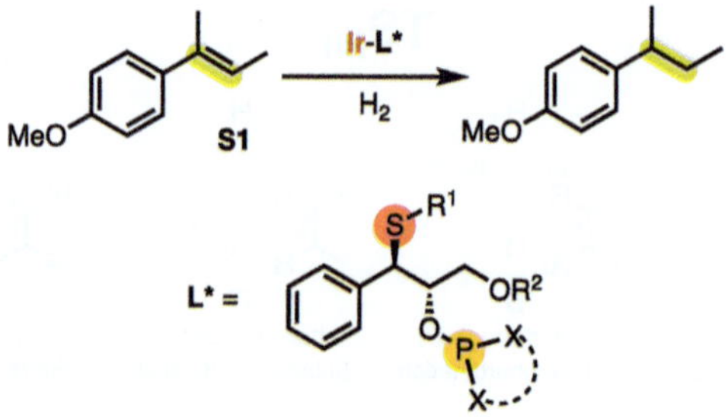

Scheme 3 Ir-catalyzed asymmetric hydrogenation of **S1** with thioether-phosphite/phosphinite ligand.

which accelerates calculations, while also featuring two types of thioether groups. In both cases, the computed most stable transitions states for both major and minor isomers arose from the migratory insertion pathway (see Scheme 2). Scrutiny of their geometries shows that the aromatic substituent in the ligand backbone may play a crucial role in the stereochemical outcome, owing to its proximity to the thioether substituent and the orientation of one of its hydrogens at the *ortho* position towards the metal center (Fig. 2).

Based on these findings, calculations were performed for a new ligand, **L3**, containing a bulkier mesityl substituent on the aromatic group of the ligand backbone (Fig. 3). According to the calculations, the energy gap between the two most stable transition states increases, indicating a potential enhancement of the enantioselectivity. Combining this result with those from the preliminary experimental study, further calculations were performed for a new ligand, **L4**, wherein the diphenyl phosphinite moiety of **L3** was replaced by a di-*o*-tolyl-phosphinite group (Fig. 3). Calculations predicted that **L4** may not significantly alter enantioselectivity compared with **L3**. However, both new ligands, **L3** and **L4**, were synthesized, and as predicted by the calculations, while the enantioselectivity was higher compared to the previous ligands, it remained quite similar between them (Fig. 3).

Going a step further on the DFT-guided catalyst design the next example discussed herein is related with an in silico-driven optimization of a family of Ir/P,S-catalysts *(11)*. Initially, DFT calculations were conducted for the enantiodetermining transition states with **L5** and **L6** in combination again with the model trisubstituted olefin **S1** (Scheme 4). Calculations revealed that the migratory insertion pathway is predominantly favored in this catalytic cycle (see Scheme 2). Moreover, the Ir-catalyst containing ligand **L6** would provide a higher enantioselectivity compared to **L5**, which was further validated experimentally.

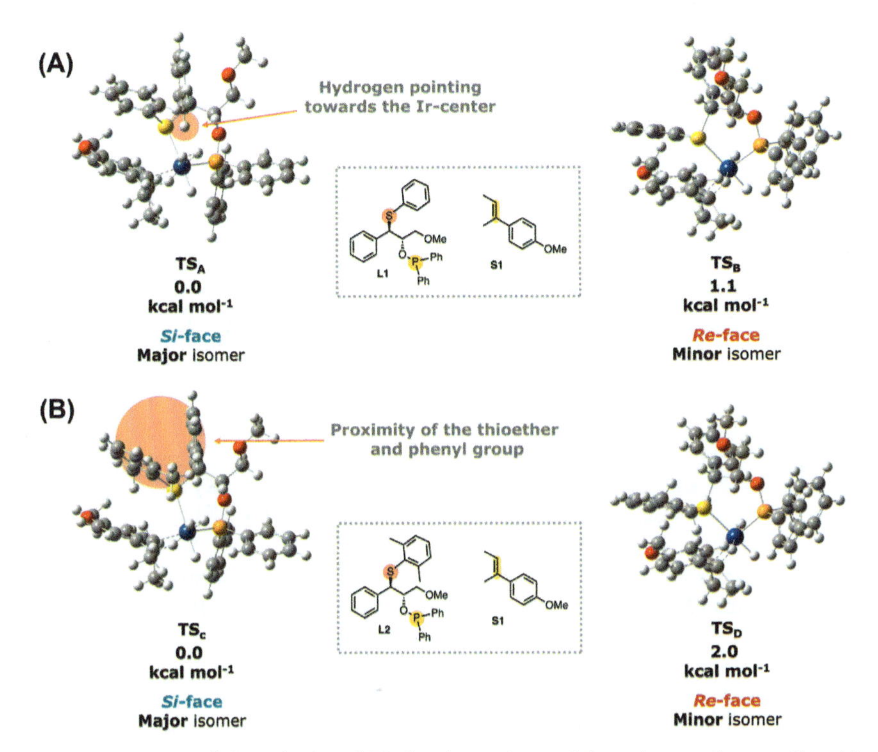

Fig. 2 Structures of the calculated TSs for the major and the minor pathways: (A) with ligand **L1**; (B) with ligand **L2**.

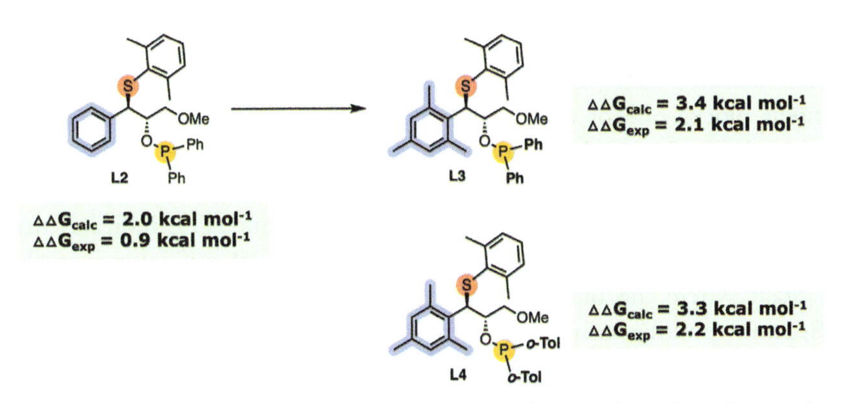

Fig. 3 Calculated and experimental Gibbs free energies for **L2** and new ligands **L3** and **L4**.

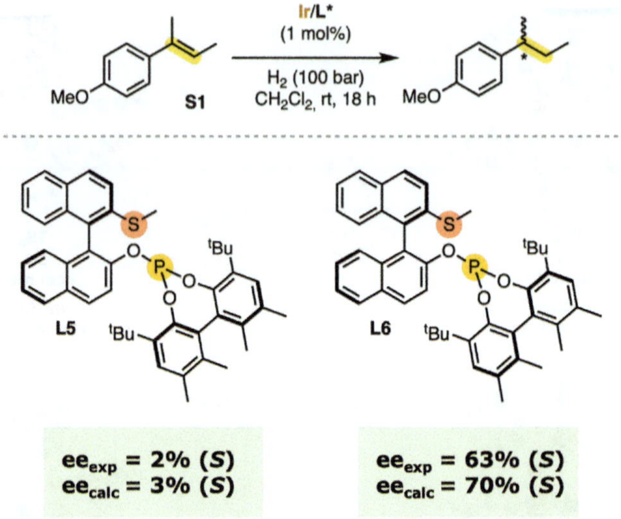

Ir/L* (1 mol%)
H_2 (100 bar)
CH_2Cl_2, rt, 18 h
MeO S1
MeO

L5
L6

ee_{exp} = 2% (*S*)
ee_{calc} = 3% (*S*)

ee_{exp} = 63% (*S*)
ee_{calc} = 70% (*S*)

Scheme 4 Ir-catalyzed asymmetric hydrogenation of model trisubstituted substrate **S1** with ligands **L5** and **L6** with the values of experimental and calculated enantioselectivities.

After validation, the most stable transition states were analyzed by the quadrant model diagram representation using MolQuO software. Careful analysis of these quadrant model diagrams revealed that the thioether substituent of the ligand plays a crucial role, and the configuration of the phosphorus moiety exhibits a match/mismatch effect. Specifically, ligand **L6**, with an (*R*)-configuration on the biphenyl phosphite moiety, represents the match combination.

Fig. 4A presents the quadrant model representation of the two most stable transition states for Ir/**L6**. In this model, the methyl thioether substituent blocks one of the quadrants, making it the most hindered and forcing the olefinic hydrogen to occupy this position in the most stable TS, $\mathbf{TS_E}$. In contrast, in $\mathbf{TS_F}$, that leads to the opposite enantiomer, the olefinic aryl group is located in this same hindered quadrant. This indicates that further modifications to the ligand that increase the blocking of the quadrant where the olefinic aryl group is located would destabilize the $\mathbf{TS_F}$, thus resulting in higher enantioselectivities.

Consequently, the quadrant model diagrams were recalculated with three new ligands, **L7–L9**, featuring bulkier thioether substituents (Fig. 4B). As expected, the bulkier the thioether substituent, the more hindered the quadrant containing this substituent, potentially resulting in higher enantioselectivity.

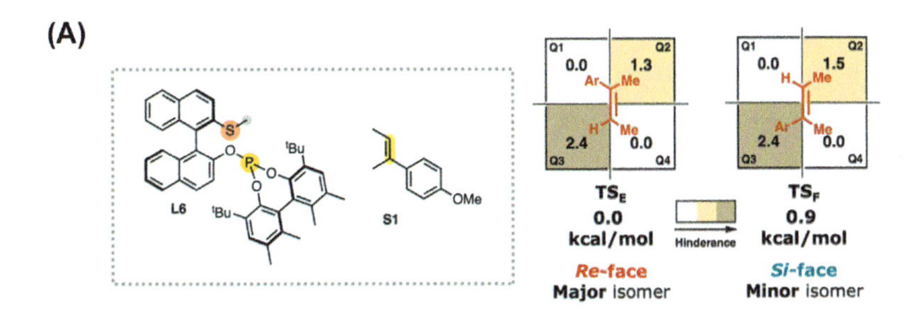

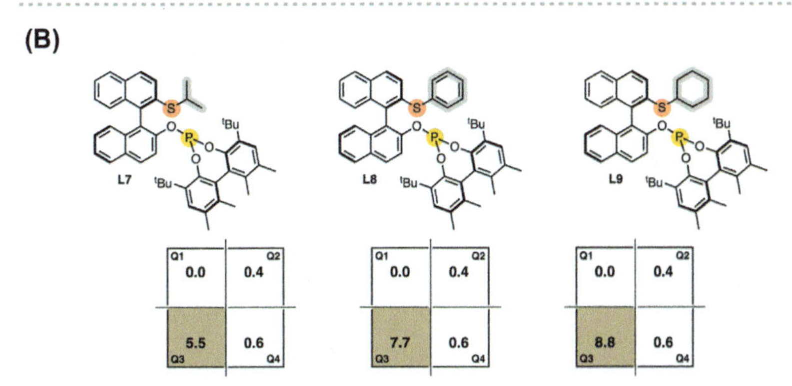

Fig. 4 Quadrant model diagram representations: (A) Most stable transition states of Ir/**L6** with olefin **S1**; (B) Most stable transition state with Ir/**L7**–**L9** without olefin **S1**.

Following these calculations, two of the three new ligands were synthesized, **L7** and **L9**. Consistent with the predicted results from the quadrant occupancy, the enantioselectivity increased with the bulkiness of the alkyl thioether substituent. Therefore, ligand **L9** was identified as the best performer and was successfully applied to a set of olefins with the same substitution pattern and olefin geometry than **S1**.

The insights gained from the quadrant model diagram not only can contribute to speed up the design of more effective catalysts, but also can offer information on the most suitable substrates to hydrogenate, as discussed in the next example involving a family of aminophosphine-oxazoline ligands (**L10**–**L13**, Fig. 5A). The related Ir-catalysts precursors provide excellent results in the asymmetric hydrogenation of a large number of di-, tri- and tetrasubstituted olefins (ee's up to 99%) *(12)*.

After initial screening on benchmark tri- and tetrasubstituted olefins (**S1** and **S2**, respectively, Fig. 5B), Ir/**L10b** and Ir/**L13c** were identified as the most effective catalysts. Ir/**L13c** with a ^{t}Bu group on the oxazoline moiety exhibit enhanced performance in the reduction of trisubstituted

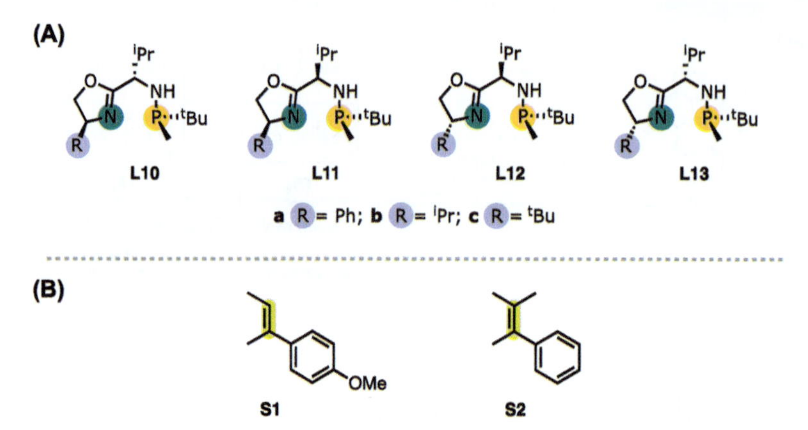

Fig. 5 (A) Aminophosphine-oxazoline ligands **L10–L13**; (B) Model tri- and tetra-substituted alkenes **S1** and **S2** used in the DFT study.

alkenes. Conversely, for tetrasubstituted olefins, Ir/**L10b** catalyst featuring a less bulky ^{i}Pr substituent yields superior performance. Similarly, the impact of the diastereomeric backbone varies between the trisubstituted alkene and the tetrasubstituted olefin.

To investigate the underlying reasons for this different behavior, DFT calculations were conducted using Ir/**L10b** and Ir/**L13c**. In line with previous studies, in all cases the computational results indicated that both enantiomers of the substrate arise from the migratory insertion pathway (**3/5-MI**, Scheme 2). In this instance, the origins of enantioselectivity were analyzed by quadrant model representations utilizing MolQuO software.

Fig. 6A shows the quadrant model representation of the two most stable transition states for the major and minor isomers with Ir/**L13c** and the benchmark trisubstituted olefin where the oxazoline substituent block the lower-left quadrant, while the methylenic carbon of the oxazoline partly occupies the upper-left quadrant. The other two quadrants are empty. According to this model, the olefin prefers the coordination through the *Re*-face, locating the smallest substituent in the most hindered quadrant and the aryl group in the semi-hindered quadrant. In contrast, the coordination through the *Si*-face, situates the aryl group at the most hindered quadrant resulting in a less favorable TS. Therefore, the model indicates that the stereochemical outcome with trisubstituted olefin depends on steric factors.

Interestingly, the analysis of this model allows the prediction of the most suitable substrates for these Ir-catalysts precursors. The model indicates that replacing the aryl group with a bulkier substituent could destabilize $\mathbf{TS_H}$, consequently leading to an enhanced enantioselectivity. This prediction was confirmed through the hydrogenation of the purely alkyl substrates with bulkier cyclohexyl and isopropyl substituents, **S3** and **S4**, respectively (Fig. 6B).

The calculations involving the tetrasubstituted olefin were performed with Ir/**L10b** and Ir/**L13c** catalysts, and provided different sense and values of enantioselectivity, 85% ee (*S*) for Ir/**L10b** and 31% ee (*R*) for Ir/**L13c**, (Scheme 5A). These catalysts present opposite configuration in the oxazoline substituent, resulting in a shift of the most hindered quadrant and consequently, in the preferred coordination face of the olefin and therefore in the resulting enantiomer, as shown in Scheme 5A. Nonetheless, the calculated value of enantioselectivity is overestimated with Ir/**L13c**. This issue was addressed conducting deuterium labeling experiments, which revealed the existence of a competitive isomerization process with Ir/**L13c** (Scheme 5B). These findings underscore the importance of combining experimental and computational techniques, which in this instance offer a comprehensive understanding of the factors governing the enantioselectivity.

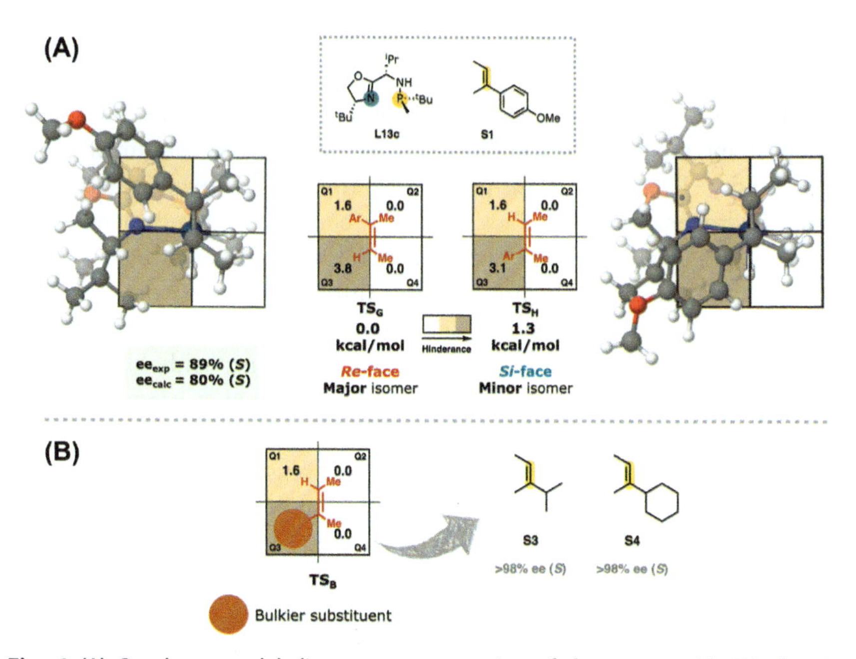

Fig. 6 (A) Quadrant model diagram representation of the most stable TSs for the asymmetric hydrogenation of **S1** with Ir/**L13c**; (B) Outcomes of the enantioselective hydrogenation for the predicted suitable substrates **S3** and **S4**.

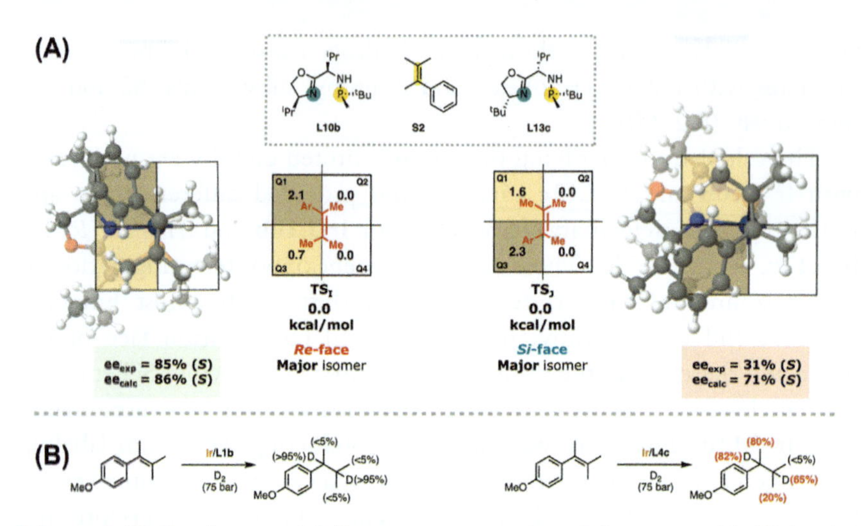

Scheme 5 (A) Quadrant model diagram representation of the most stable TS for the asymmetric hydrogenation of **S2** with Ir/**L10b** and Ir/**L13c**; (B) Deuterium labeling experiments.

As evidenced by prior examples, steric factors predominantly control the enantioselectivity in the Ir-catalyzed asymmetric hydrogenation of non-chelating olefins. However, there are exceptions, such as the subsequent and final example related with phosphine-triazole ligands, where the quadrant model representation fails to elucidate the origin of enantioselectivity. These Ir-catalysts precursors provide high enantioselectivities in the hydrogenation of various exocyclic benzofused-based olefins (ee's between 92% and 99%) *(9)*. These olefins pose a challenge due to their difficult hydrogenation which is primarily attributed to their propensity to isomerize into the corresponding endocyclic olefins under hydrogenation conditions. Nonetheless, deuterogenation experiments with four-, five-, six- and seven-membered benzofused olefins showed deuterium incorporation solely in the olefinic carbons, indicating that isomerization does not occur irrespective of the size of the benzofused ring. This unequivocally demonstrates that enantioselectivity is predominantly governed by the constraints of the catalyst's chiral pocket. Therefore, the transition states responsible for the enantioselectivity were studied through DFT calculations using the best-performing ligand **L14** and model exocyclic substrate **S5** (Scheme 6). The most stable transition states emerged once more from the migratory insertion pathway (see Scheme 2).

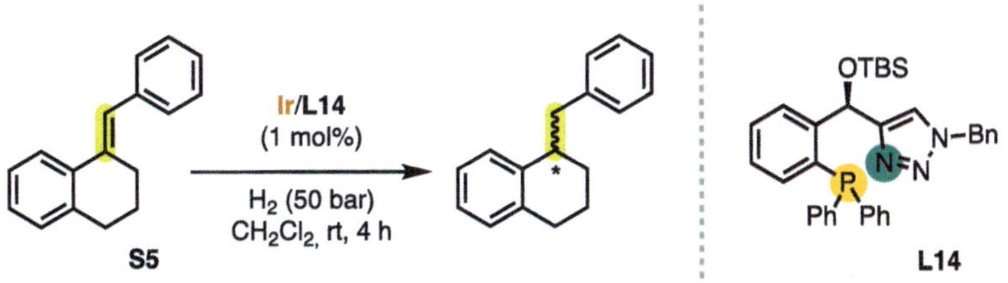

Scheme 6 Ir-catalyzed asymmetric hydrogenation of model exocyclic substrate **S5** with ligands **L14**.

Analysis of the geometries of the most stable transition states revealed a series of attractive interactions between the substrate and the ligand, as shown in the NCI plots (Fig. 7). For $\mathbf{TS_K}$, three CH–N, two CH-π, and one T-shaped π–π stabilizing interactions stabilize were found. In contrast, $\mathbf{TS_L}$ is stabilized by only two CH–N and one T-shaped π–π interaction. These attractive interactions make the cavity of $\mathbf{TS_K}$ better suited for olefin **S5** than the chiral pocket of $\mathbf{TS_L}$, which lacks some of these stabilizing interactions.

It should be pointed out that the stereochemical model also highlighted the importance of the triazole and TBS groups in the model design, as demonstrated by the CH–N interactions and the two CH–π interactions involving the TBS group in $\mathbf{TS_K}$. However, these interactions depend on the relative position of the tetrahydronaphthalene phenyl ring, which is influenced by the ring size of the benzofused moiety, correlating with the variations in enantiomeric excesses observed with different ring sizes. While the chiral pocket accommodates exocyclic olefins with five-, six- and seven-membered benzofused rings well, the four-membered exocyclic substrate shifts the tetrahydronaphthalene phenyl ring, leading to the loss of these key interactions.

Through the analysis of these examples, we can conclude that the migratory insertion pathway is the most probable followed in this catalytic cycle. Nevertheless, as in the case of previous studies, there was not enough difference found with the σ-bond metathesis pathway to directly discard it in future calculations. Additionally, steric factors predominantly dictate the enantioselectivity, and thus the use of the quadrant model can be optimal for predicting new catalysts or suitable substrates. However, it is important to also consider other factors that, while less frequent, can play a crucial role in the stereochemical outcome.

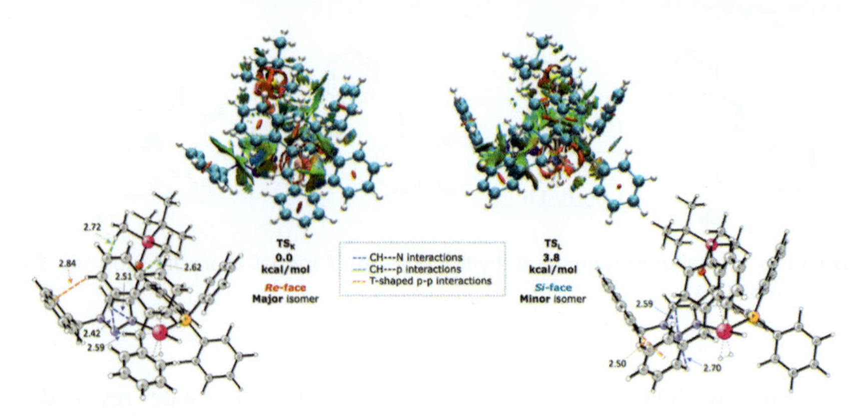

Fig. 7 Structures of the most stable transition states for Ir/**L14** with **S5** and their NCI-plots. Strong and attractive interactions are blue, weak interactions are green, and strong and repulsive interactions are red.

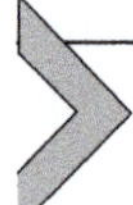

4. Pd-catalyzed asymmetric allylic substitution reaction

One of the most critical challenges in the field of organic synthesis is the formation of stereoselective carbon-carbon (C–C), carbon-nitrogen (C–N), and carbon-oxygen (C–O) bonds. Achieving high levels of stereoselectivity in these bond-forming reactions is essential for the synthesis of complex molecules.

Among the various methodologies developed to address this challenge, palladium-catalyzed asymmetric allylic substitution has emerged as a particularly effective and versatile strategy. This approach offers several significant advantages that make it highly attractive for synthetic chemists. This process typically operates under mild reaction conditions and exhibits broad functional group tolerance *(1–4,57–69)*.

4.1 Overview of mechanistic aspects

The mechanism for the Pd-catalyzed asymmetric allylic substitution reaction has been extensively studied and its well established (Scheme 7) *(60–72)*. The catalytic cycle starts with the coordination of the allylic substrate to the catalytic active specie, $\mathbf{Int_1}$, to form $\mathbf{Int_2}$ (Scheme 7). The cycle continues with the oxidative addition of $\mathbf{Int_2}$, yielding the π-allyl intermediate ($\mathbf{Int_3}$, Scheme 7). Next, a nucleophilic attack takes place to obtain $\mathbf{Int_4}$, an unstable Pd^0 olefin complex. This latter intermediate undergoes final product decoordination and release.

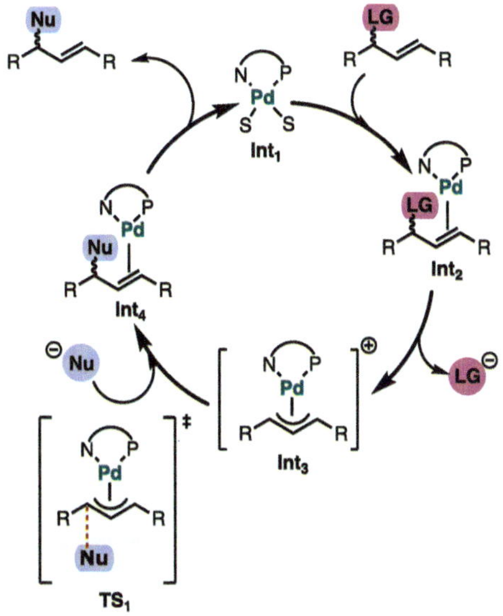

Scheme 7 Catalytic cycle for Pd-catalyzed asymmetric allylic substitution reaction. S = solvent or vacant; LG = leaving group; Nu = nucleophile.

For this process, it was found that the rate-determining step can be either the oxidative addition or the nucleophilic attack *(57–69)*. When the benchmark symmetric substrate examined in this book chapter is used, it is accepted that the nucleophilic attack is the rate determining step. This attack occurs through an outer-sphere mechanism on one of the two allylic terminal atoms (**TS$_1$**). For such substrates, it is also known that **Int$_2$** can primarily undergo interconversion via the well-stablished π-σ-π isomerization mechanism between the two most stable *syn/syn* isomers *(57–69)*. Consequently, the stereochemical outcome depends on the ability of the chiral catalyst to differentiate between the two allylic terminal carbons and the two *syn/syn* isomers.

Hence, to calculate/predict the stereochemical result through DFT calculations, it's necessary to compute both *syn/syn* isomers and the TS for the nucleophilic attack. On top of that, in the case of for P,S-ligands two possibilities have to be taken into account, *trans* to both P and S heteroatoms. In the case of P,N-ligands, solely the TS *trans* to the phosphorous needs to be considered, due to widely acknowledged larger reactivity of the allyl *trans* to a P atom in front of the *trans* to a N atom (Fig. 8). Furthermore, a conformational search of the chelating ring of all these key isomers is crucial for each individual case.

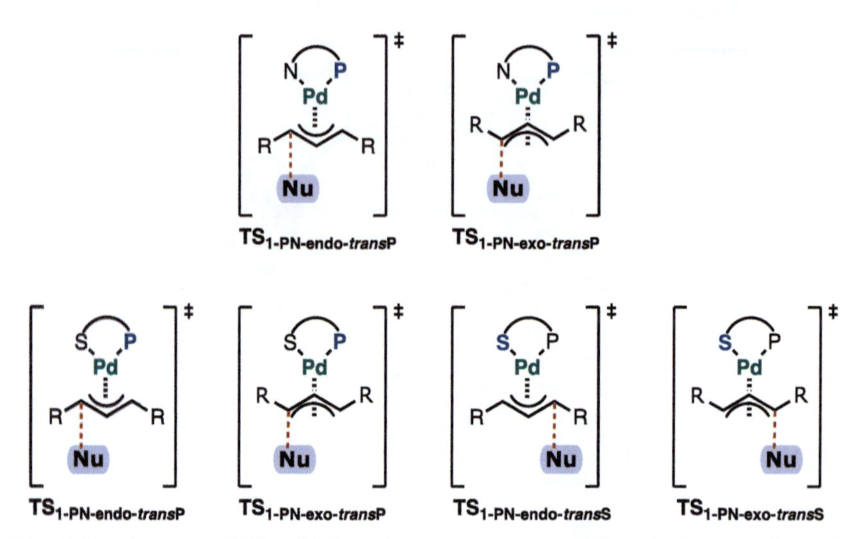

Fig. 8 Key isomers of $\mathbf{TS_1}$ of Pd-catalyzed asymmetric allylic substitution of benchmark symmetric substrates for P,N-ligands and P,S-ligands.

Additionally, it should be noted that in all the calculations of this book chapter ammonia was used as nucleophile. Using ammonia speed up the calculations and circumvents the issues associated with charge separation along with continuum solvent model when using malonates as nucleophiles *(73,74)*.

4.2 Representative examples

This section begins with an example of the application of a family of phosphite-oxazoline ligands in Pd-catalyzed asymmetric allylic substitution. This ligand library has been applied to numerous hindered and unhindered substrates with a wide range of C-, O-, and N-nucleophiles. By carefully selecting the ligand parameters, high activities and excellent enantioselectivities (ee's up to 99%) have been achieved *(13)*.

In this study, DFT and NMR analyses of the key Pd-allyl complexes were conducted to better understand the origins of the excellent enantioselectivities observed experimentally. In this sense, calculations were performed on the TSs responsible of the enantioselectivity with **L15** and **L16** with benchmark hindered and unhindered symmetric substrates, **S6** and **S7**, respectively (Scheme 8). The study of these two ligands provided insights into the effect of the configuration of the biaryl phosphite moiety.

Fig. 9 shows the two most stable transition states leading to opposite enantiomers for substrates **S6** using **L15** ($\mathbf{TS_M}$ and $\mathbf{TS_L}$) and **L16** ($\mathbf{TS_O}$ and $\mathbf{TS_P}$), along with their respective NCI plots.

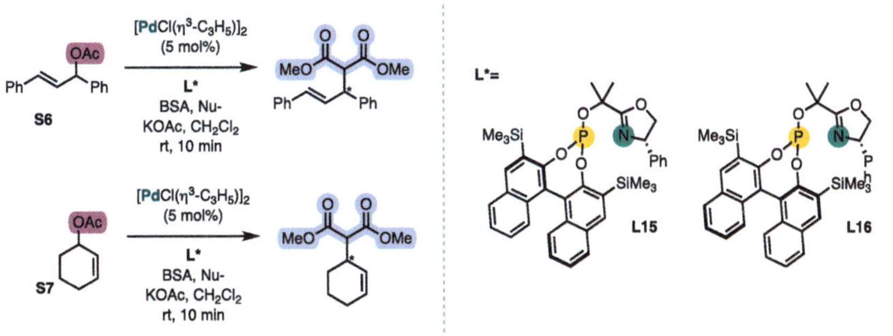

Scheme 8 Pd-catalyzed asymmetric allylic substitution of benchmark hindered and unhindered symmetric substrates, **S6** and **S7**, respectively, using ligand **L15** and **L16**.

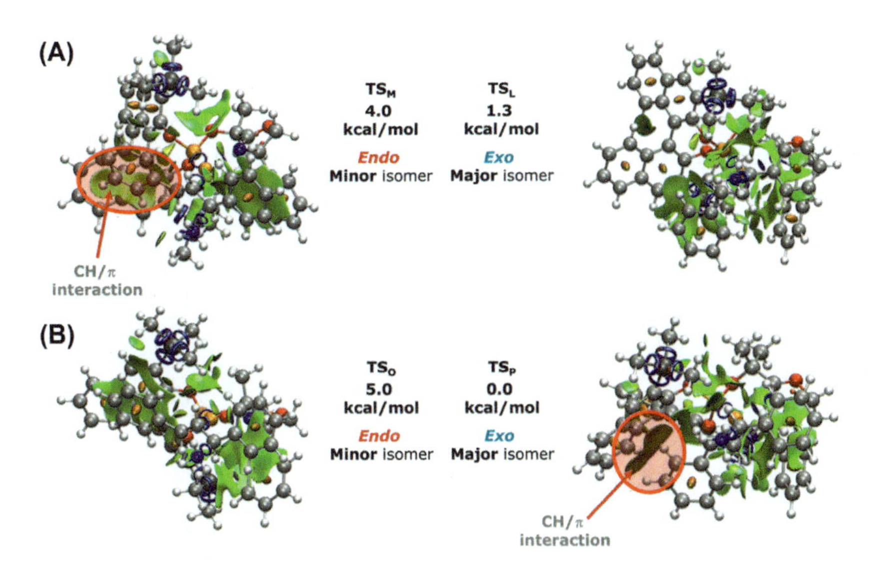

Fig. 9 NCI plots of the most stable TSs from **S6** using ligands (a) **L15** and (b) **L16**. Strong and attractive interactions are blue, weak interactions are green, and strong and repulsive interactions are red.

Analysis of these geometries reveals that in both catalytic systems, the *endo* TSs are destabilized due to a steric repulsion generated between one of the phenyl substituents of the substrate and the oxazoline substituent. This unfavorable interaction explains why the same product configuration is achieved with both ligands. Notably, for Pd/**L16**, the *exo* TS presents a CH/π interaction between one of the phenyl rings of the substrate and the biaryl phosphite group, further stabilizing this TS. This stabilizing interaction increases the energy gap between the *endo* and *exo* TSs, accounting for the higher enantioselectivity observed with Pd/**L16**.

In contrast, for substrate **S7**, the NCI plots of the two most stable TSs with ligands **L15** and **L16** show weak stabilizing attractive interactions between the substrate and one of the aryls of the phosphite group (Fig. 10). However, with ligand **L15**, this favorable interaction occurs in the *exo* TS, leading to the (*R*)-product. With ligand **L16**, this stabilizing interaction is found in the *endo* TS responsible for the (*S*)-product. This explains the obtention of opposite enantiomers when using these ligands. Furthermore, these weak attractive interactions are stronger in the *endo* TS of ligand **L16** than in the *exo* TS of ligand **L15**, correlating with the higher enantioselectivity observed with ligand **L16**.

To gain a complete understanding of the ligand design parameters that influence enantioselectivity, NMR studies were conducted using ligands **L17** and **L18** to elucidate the role of the substituents in the ligand's backbone chain. The variable-temperature (VT) NMR study (30 °C−−80 °C) of Pd-1,3-diphenyl allyl complexes **1** and **2** revealed a mixture of the two *syn/syn* isomers in equilibrium at ratios 3:1 and 10:1, respectively (Scheme 9A). In situ NMR analysis of the reactivity of complexes **1** and **2** with sodium dimethyl malonate indicated that both *syn/syn* isomers react with comparable rates. Given the similar reaction rates, the higher enantioselectivity achieved

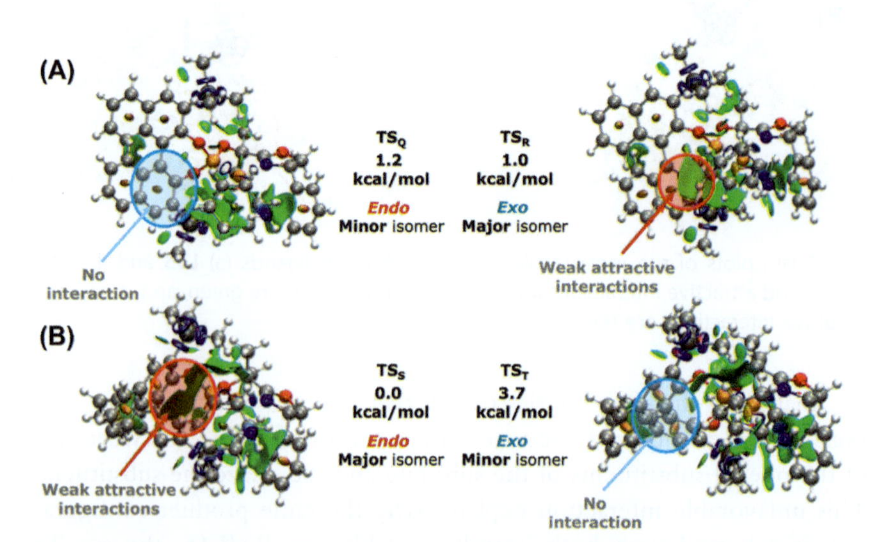

Fig. 10 NCI plots of the most stable calculated TSs (TS(R) endo and TS(S) exo) from **S7** using ligands (A) **L15** and (B) **L16**. Strong and attractive interactions are blue, weak interactions are green, and strong and repulsive interactions are red.

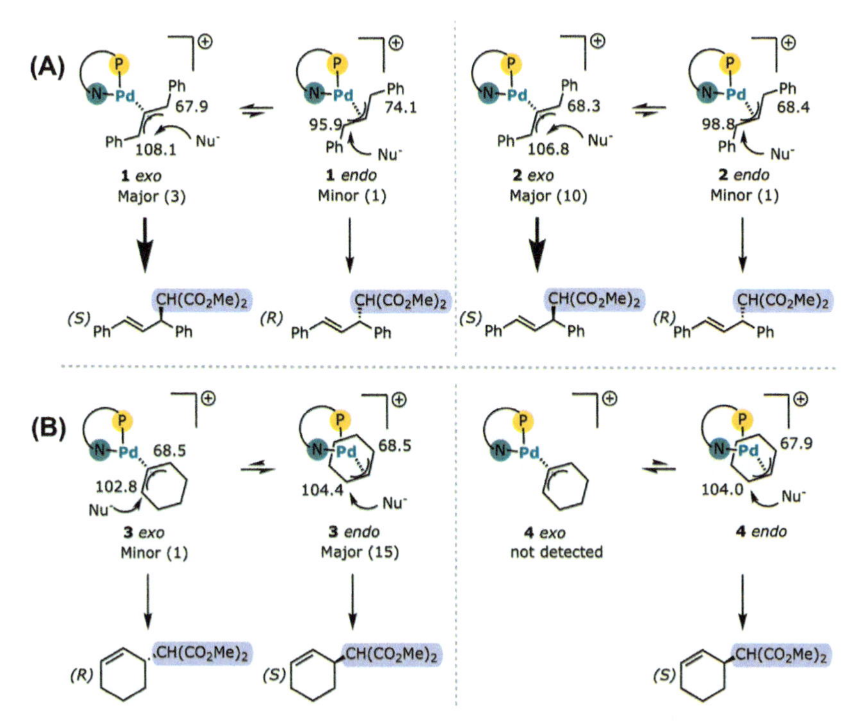

Scheme 9 (A) Diasterisomeric Pd-allyl intermediates for **S6** with **L17** and **L18** (B) Diasterisomeric Pd-allyl intermediates for **S7** with **L17** and **L18**.

with Pd/**L18** is attributed to the higher relative population of the faster-reacting isomer, which indicate that the ability of this catalytic system to control both the population and relative electrophilicity of the Pd-allyl intermediates is crucial for achieving excellent enantiocontrol.

The VT NMR study of the Pd-1,3-cyclohexenyl allyl complex **3** displayed a mixture of the two *syn/syn* isomers in fast equilibrium in a 15:1 ratio, whereas intermediate **4** showed only one isomer (Scheme 9B). For complex **3**, the electrophilicity of the allylic carbon *trans* to the P is very similar in the *endo* and *exo* isomers, suggesting that both isomers should react at comparable rates. Consequently, enantioselectivity is primarily governed by the relative ratio of *endo* and *exo* isomers. The higher enantioselectivity observed with Pd/**L18** may be associated to the detection of only the endo isomer.

In summary, DFT calculations in combination with the NMR studies indicate that while the enantioselectivity for cyclic substrates is primarily controlled by the biaryl phosphite groups and the substituents on the alkyl backbone chain, for linear substrates, the oxazoline substituent also plays a crucial role.

After delving into this interesting example of rationalization of the factors governing enantioselectivity through a combination of theoretical and experimental techniques, we will advance our discussion with two examples about the prediction of new catalysts for the Pd-catalyzed asymmetric allylic substitution reaction. This next step involves integrating experimental results with DFT calculations to catalyst design with enhanced performance.

The first example involves the computationally guided design of a family of phosphite-thioether ligands (Scheme 10) *(14)*. Initial screening of these ligands in the Pd-catalyzed asymmetric allylic substitution revealed that the best enantioselectivities for model linear substrate **S6** (ee's up to 97%) were achieved with ligands containing a combination of an aryl thioether group and an (*R*)-biaryl phosphite group. For benchmark substrate cyclic **S7**, the aryl thioether substituent is also important, but both configurations of the biaryl phosphite moiety provided good results, with ee's up to 88%.

In the DFT study, **L19** and **L20** were used to evaluate the impact of the chiral axis of the biaryl phosphite moiety (Fig. 11A). Theoretical calculations correctly reproduced the experimental outcome with both ligands. However, directly analyzing the two most stable TSs proves challenging in identifying the factors influencing enantioselectivity. Therefore, a quadrant analysis was conducted (Fig. 11A). The critical factor influencing selectivity lies in the steric repulsion between one of the phenyl substituents of the substrate and the biaryl phosphite moiety. Additionally, the steric bulk on the "sulfur side" of the catalyst also plays a crucial role, pushing the sterically active regions of the substrate and catalyst closer together, thereby enhancing selectivity.

Further modifications on the "sulfur side" were subsequently recalculated to explore potential improvements in enantioselectivity. Thus, calculations were performed with bulkier thioether substituents leading to ligands **L21**

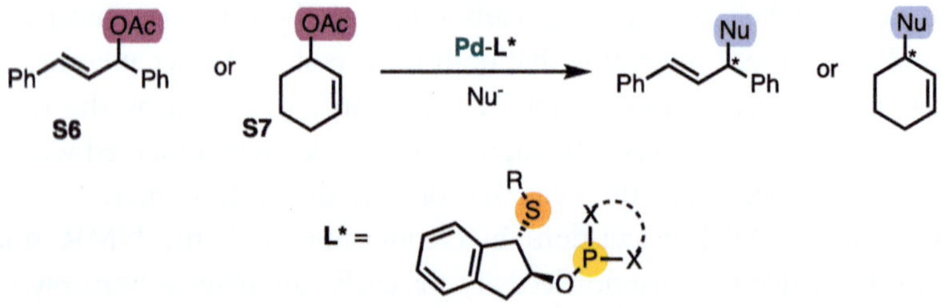

Scheme 10 Pd-catalyzed asymmetric allylic substitution of **S6** and **S7** with the thioether-phosphite ligand library.

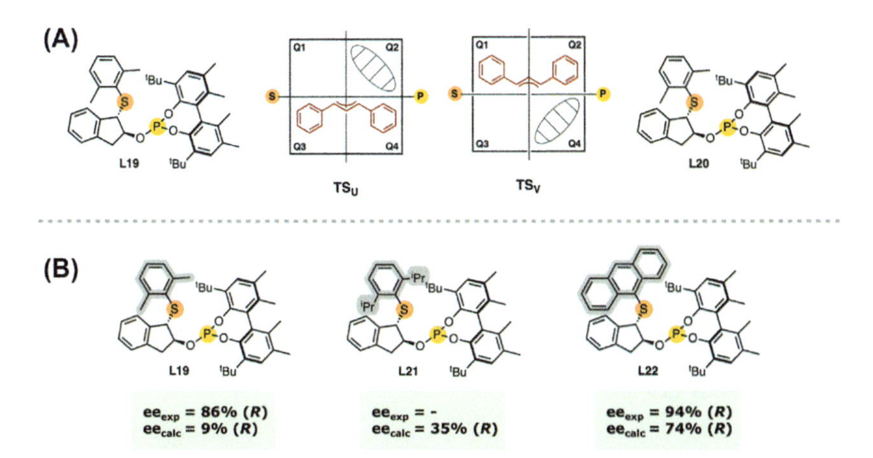

Fig. 11 (A) Schematic representation of the quadrant model of the most stable TS with **L19** and **L20** using **S6**; (B) Calculated and experimental enantioselectivities for **L20–L22** using **S7**.

and **L22**, incorporating 2,6-diisopropylphenyl or anthryl moieties (Fig. 11B). These calculations focused on cyclic substrate **S7**, which exhibited lower enantioselectivity. The results show that enantioselectivity increases with the steric bulk of the thioether substituent, therefore ligand **L22**, containing an anthryl thioether group was synthetized and tested in the Pd-catalyzed alkylation of **S7**. The enantioselectivity increased from 86% ee to 94% ee, consistent with theoretical predictions (Fig. 11B).

Theoretical calculations facilitated a more efficient and cost-effective process, identifying ligand **L22** as optimal. Subsequently, this new ligand was applied across various cyclic substrates in combination with a diverse range of nucleophiles, achieving exceptional levels of enantioselectivity.

Visual examination of the structure can provide important insights into the factors governing enantioselectivity, but it is not always trivial as demonstrated in the latter example. Nonetheless, in all cases, recalculating the TSs of the proposed modifications is always crucial, as unexpected results can sometimes arise. This is particularly evident in the following case involving a family of amino-phosphite ligands.

Initially, a small set of amino-phosphite ligands were tested on the Pd-catalyzed asymmetric allylic substitution of benchmarks substrates **S6** and **S7** *(15)*. The best results for **S6** were obtained with ligand **L23**, while for **S7** were achieved using **L24**. Therefore, these ligands were used in the

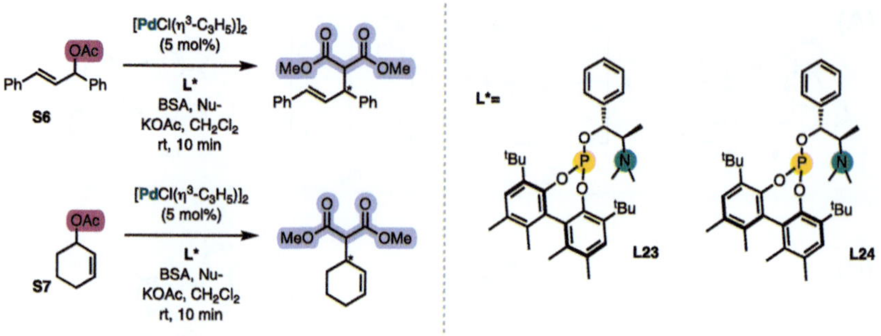

Fig. 12 Pd-catalyzed asymmetric allylic substitution of **S6** and **S7** with the amino-phosphite ligands **L23** and **L24**.

DFT study along with **S6**, to identify which ligand parameters should be further modified to enhance the enantioselectivity (Fig. 12).

The analysis of the TSs structures for both the major and minor pathways with each ligand revealed that, in all TSs, the methyl substituent of the ephedrine backbone points away from the coordination sphere (Fig. 13A). This orientation suggests that the methyl substituent has little impact on enantioselectivity. Nonetheless, recalculating the TSs for **L25** (Fig. 13B), without the methyl substituent, provided a bigger energy gap between the two TSs leading to opposite enantiomers of the alkylated product than **L23** and **L24**, indicating that **L25** should provide higher enantioselectivity. Another modification of the ligand, **L26**, which involved switching the position of the phenyl substituent, was also studied using DFT calculations (Fig. 13B). However, the theoretical calculations did not predict any improvement in enantioselectivity.

Both modified ligands were subsequently synthesized and tested. As anticipated by the DFT calculations, the ligand without the methyl substituent (**L25**) provided the highest enantioselectivities in the allylic alkylation of **S6** and **S7**, while the use of **L26** resulted in enantioselectivities similar to those previously obtained with **L24**.

Therefore, once again, thanks to DFT calculations, ligand **L23** was identified as a highly effective catalyst. This ligand was subsequently applied across a wide range of substrates and nucleophiles, consistently yielding excellent enantioselectivities. The success of **L23** underscores the power of computational methods in streamlining the discovery and optimization of catalysts.

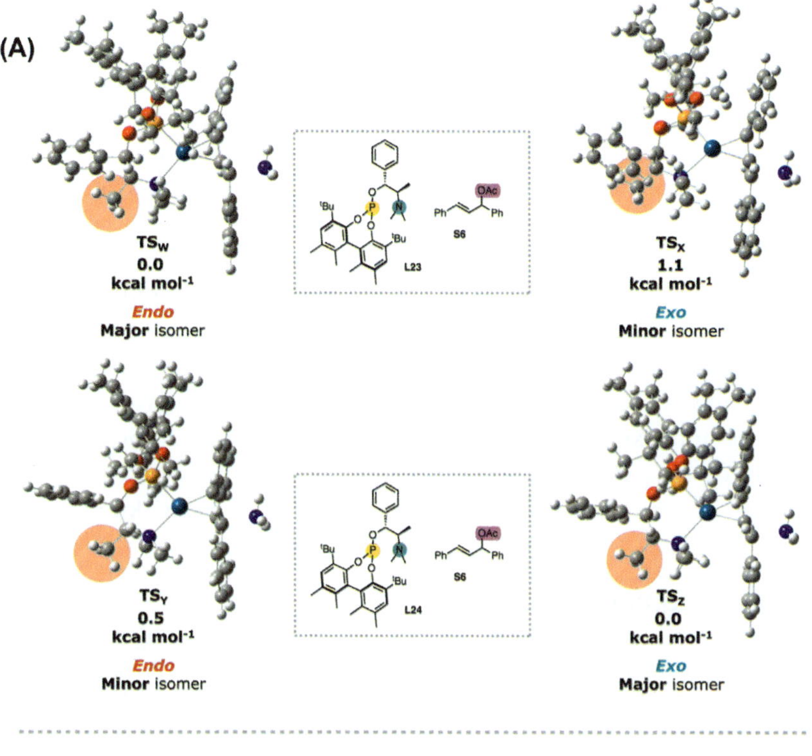

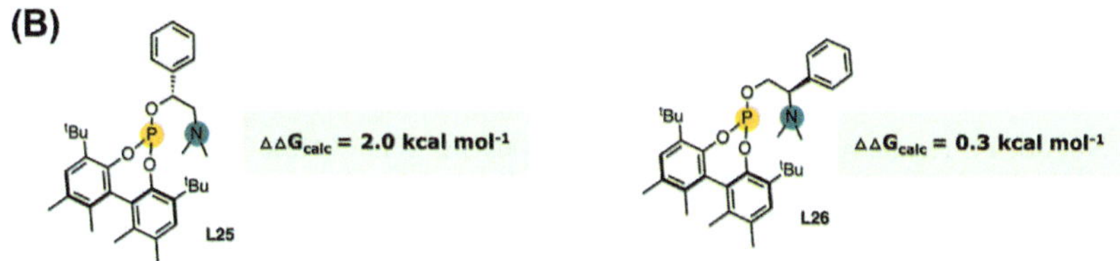

Fig. 13 (A) Structures of the calculated TSs for the major and the minor pathways with ligands **L23** and **L24**; (B) Calculated and experimental enantioselectivities with **L20–L22** using **S7**.

5. Conclusions

In summary, in this chapter, we have showcased various examples of how DFT calculations has helped our research group understanding the factors governing the enantioselectivity in two asymmetric metal-catalyzed reactions: Ir-hydrogenation of non-chelating olefins and Pd-allylic substitution reactions. Through analyzing the most stable transition states leading to opposite enantiomers, we have demonstrated how DFT calculations can speed up catalyst design. By leveraging the insights gained from both approaches, we can systematically explore and optimize catalyst structures, paving the way for significant advancements in the field of asymmetric synthesis. Moreover, through the combination of theoretical and experimental methods, we have uncovered key factors influencing enantioselectivity in these reactions.

The predictive power of DFT has allowed us to move beyond trial and error, accurately anticipating reaction outcomes. This not only speeds up catalyst discovery but also leads experimentalists to explore new avenues where intuition alone might fall short.

Looking ahead, the integration of machine learning with DFT calculations and experimental studies holds great promise for advancing catalyst design. By embracing machine learning's potential, we are entering a new era where innovation in catalyst design knows fewer constraints.

Acknowledgments

We thanks all co-workers and collaborators who contributed to this work, in particular to Jorge Faiges, Dr. Jéssica Margalef and Prof. Per-Ola Norrby. We also thank to FEDER/ Ministerio de Ciencia e Innovación (MICINN)/AEI for grants PID2022–139996NB-I00 and PID2021–128128NB-100 to the Catalan Government for grant 2021SGR00163. M. B. also thanks Ministerio de Ciencia e Innovación for a Juan de la Cierva fellowship.

References

1. Noyori, R. *Asymmetric Catalysis in Organic Synthesis;* Wiley: New York, 1994.
2. Jacobsen, E. N., Pfaltz, A., Yamamoto, H., Eds. *Comprehensive Asymmetric Catalysis.* Springer-Verlag: Berlin, 1999.
3. Blaser, H.-U., Federsel, H.-J., Eds. *Asymmetric Catalysis in Industrial Scale: Challenges, Approaches and Solutions.* 2nd ed., Wiley: Weinheim, 2010.
4. Akiyama, T., Ojima, I., Eds. *Catalytic Asymmetric Synthesis.* 4th ed. John Wiley & Sons, Inc: Hoboken, 2022.
5. Sameera, W. M. C.; Maseras, F. Transition Metal Catalysis by Density Functional Theory and Density Functional Theory/Molecular Mechanics. *WIREs Comput. Mol. Sci.* **2012,** *2*, 375–385.
6. Sperger, T.; Sanhueza, I. A.; Kalvet, I.; Schoenebeck, F. Computational Studies of Synthetically Relevant Homogeneous Organometallic Catalysis Involving Ni, Pd, Ir, and Rh: An Overview of Commonly Employed DFT Methods and Mechanistic Insights. *Chem. Rev.* **2015,** *115*, 9532–9586.
7. Lam, Y.-H.; Grayson, M. N.; Holland, M. C.; Simon, A.; Houk, K. N. Theory and Modeling of Asymmetric Catalytic Reactions. *Acc. Chem. Res.* **2016,** *49*, 750–762.
8. Santoro, S.; Kalek, M.; Huang, G.; Himo, F. Elucidation of Mechanisms and Selectivities of Metal-Catalyzed Reactions using Quantum Chemical Methodology. *Acc. Chem. Res.* **2016,** *49*, 1006–1018.

9. Biosca, M.; Cruz-Sánchez, P.; Tarr, D.; Llanes, P.; Karlsson, E. A.; Margalef, J.; Pàmies, O.; Pericàs, M. A.; Diéguez, M. Filling the Gaps in the Challenging Asymmetric Hydrogenation of Exocyclic Benzofused Alkenes with Ir−P,N Catalysts. *Adv. Synth. Catal.* **2023,** *365*, 167–177.
10. Margalef, J.; Caldentey, X.; Karlsson, E. A.; Coll, M.; Mazuela, J.; Pàmies, O.; Diéguez, M.; Pericàs, M. A. A Theoretically-guided Optimization of a New Family of Modular P,S-Ligands for Iridium-Catalyzed Hydrogenation of Minimally Functionalized Olefins. *Chem. Eur. J.* **2014,** *20*, 12201–12214.
11. Faiges, J.; Borràs, C.; Pastor, I. M.; Pàmies, O.; Besora, M.; Diéguez, M. Density Functional Theory-Inspired Design of Ir/P,S-Catalysts for Asymmetric Hydrogenation of Olefins. *Organometallics* **2021,** *40*, 3424–3435.
12. Biosca, M.; De La Cruz-Sánchez, P.; Faiges, J.; Margalef, J.; Salomó, E.; Riera, A.; Verdaguer, X.; Ferré, J.; Maseras, F.; Besora, M.; Pàmies, O.; Diéguez, M. P-Stereogenic Ir-MaxPHOX: A Step toward Privileged Catalysts for Asymmetric Hydrogenation of Nonchelating Olefins. *ACS Catal.* **2023,** *13*, 3020–3035.
13. Biosca, M.; Saltó, J.; Magre, M.; Norrby, P.-O.; Pàmies, O.; Diéguez, M. An Improved Class of Phosphite-Oxazoline Ligands for Pd-Catalyzed Allylic Substitution Reactions. *ACS Catal.* **2019,** *9*, 6033–6048.
14. Biosca, M.; Margalef, J.; Caldentey, X.; Besora, M.; Rodríguez-Escrich, C.; Saltó, J.; Cambeiro, X. C.; Maseras, F.; Pàmies, O.; Diéguez, M.; Pericàs, M. A. Computationally Guided Design of a Readily Assembled Phosphite–Thioether Ligand for a Broad Range of Pd-Catalyzed Asymmetric Allylic Substitutions. *ACS Catal.* **2018,** *8*, 3587–3601.
15. Magre, M.; Biosca, M.; Norrby, P.-O.; Pàmies, O.; Diéguez, M. Theoretical and Experimental Optimization of a New Amino Phosphite Ligand Library for Asymmetric Palladium-Catalyzed Allylic Substitution. *ChemCatChem* **2015,** 7, 4091–4107.
16. Becke, A. D. Density-functional Thermochemistry. III. The role of Exact Exchange. *J. Chem. Phys.* **1993,** *98*, 5648–5652.
17. Stephens, P. J.; Devlin, F. J.; Chabalowski, C. F.; Frisch, M. J. Ab Initio Calculation of Vibrational Absorption and Circular Dichroism Spectra Using Density Functional Force Fields. *J. Phys. Chem. A* **1994,** *98*, 11623–11627.
18. Grimme, S.; Antony, J.; Ehrlich, S.; Krieg, H. A Consistent and Accurate Ab Initio Parametrization of Density Functional Dispersion Correction (DFT-D) for the 94 Elements H-Pu. *J. Chem. Phys.* **2010,** *132* No. 154104.
19. Hay, P. J.; Wadt, W. R. Ab Initio Effective Core Potentials for Molecular Calculations. Potentials for the Transition Metal Atoms Sc to Hg. *J. Chem. Phys.* **1985,** *82*, 270–283.
20. Hay, P. J.; Wadt, W. R. Ab Initio Effective Core Potentials for Molecular Calculations. Potentials for K to Au Including the Outermost Core Orbitals. *J. Chem. Phys* **1985,** *82*, 299–310.
21. Hehre, W. J.; Ditchfield, R.; Pople, J. A. Self-Consistent Molecular Orbital Methods. XII. Further Extensions of Gaussian—Type Basis Sets for Use in Molecular Orbital Studies of Organic Molecules. *J. Chem. Phys.* **1972,** *56*, 2257–2261.
22. Hariharan, P. C.; Pople, J. A. The Influence of Polarization Functions on Molecular Orbital Hydrogenation Energies. *Theor. Chim. Acta* **1973,** *28*, 213–222.
23. Francl, M. M.; Pietro, W. J.; Hehre, W. J.; Binkley, J. S.; Gordon, M. S.; Defrees, D. J.; Pople, J. A. Self-Consistent Molecular Orbital Methods. XXIII. A Polarization-type Basis set for Second-row Elements. *J. Chem. Phys.* **1982,** 77, 3654–3665.
24. Petersson, G. A.; Bennett, A.; Tensfeldt, T. G.; Al-Laham, M. A.; Shirley, W. A.; Mantzaris, J. A Complete Basis Set Model Chemistry. I. The Total Energies of Closed-shell Atoms and Hydrides of the First-row Atoms. *J. Chem. Phys.* **1988,** *89*, 2193–2218.
25. Tomasi, J.; Mennucci, B.; Cammi, R. Quantum Mechanical Continuum Solvation Models. *Chem. Rev.* **2005,** *105*, 2999–3094.

26. Krishnan, R.; Binkley, J. S.; Seeger, R.; Pople, J. A. Self-consistent Molecular Orbital Methods. XX. A Basis Set for Correlated Wave Functions. *J. Chem. Phys* **1980,** *72*, 650–654.
27. McLean, A. D.; Chandler, G. S. Contracted Gaussian Basis Sets for Molecular Calculations. I. Second Row Atoms, Z=11–18. *J. Chem. Phys.* **1980,** *72*, 5639–5648.
28. Besora, M.; Maseras, F. Computational Insights into Metal-catalyzed Asymmetric Hydrogenation. *Adv. Catal.* **2021,** *68*, 385–426.
29. Poater, A.; Cosenza, B.; Correa, A.; Giudice, S.; Ragone, F.; Scarano, V.; Cavallo, L.; SambVca, A. Web Application for the Calculation of the Buried Volume of N-Heterocyclic Carbene Ligands. *Eur. J. Inorg. Chem.* **2009**, 1759–1766.
30. Zuccarello, G.; Nannini, L. J.; Arroyo-Bondía, A.; Fincias, N.; Arranz, I.; Pérez-Jimeno, A. H.; Peeters, M.; Martín-Torres, I.; Sadurní, A.; García-Vázquez, V.; Wang, Y.; Kirillova, M. S.; Montesinos-Magraner, M.; Caniparoli, U.; Núñez, G. D.; Maseras, F.; Besora, M.; Escofet, I.; Echavarren, A. M. *JACS Au* **2023,** *3*, 1742–1754.
31. Aguado-Ullate, S.; Saureu, S.; Guasch, L.; Carbo, J. J. Theoretical Studies of Asymmetric Hydroformylation Using the Rh−(R,S)-BINAPHOS Catalyst—Origin of Coordination Preferences and Stereoinduction. *Chem.—Eur. J.* **2012,** *18*, 995–1005.
32. Aguado- Ullate, S.; Urbano-Cuadrado, M.; Villalba, I.; Pires, E.; García, J. I.; Bo, C.; Carbó, J. J. Predicting the Enantioselectivity of the Copper-Catalysed Cyclopropanation of Alkenes by Using Quantitative Quadrant-Diagram Representations of the Catalysts. *Chem.—Eur. J* **2012,** *18*, 14026–14036.
33. Contreras-García, J.; Johnson, E. R.; Keinan, S.; Chaudret, R.; Piquemal, J. P.; Beratan, D. N.; Yang, W. NCIPLOT: A Program for Plotting Noncovalent Interaction Regions. *J. Chem. Theory Comput.* **2011,** 7, 625–632.
34. Busacca, C. A.; Fandrick, D. R.; Song, J. J.; Senanayake, C. H. The Growing Impact of Catalysis in the Pharmaceutical Industry. *Adv. Synth. Catal.* **2011,** *353*, 1825–1864.
35. Ager, D. J.; De Vries, A. H. M.; De Vries, J. G. Asymmetric Homogeneous Hydrogenations at Scale. *Chem. Soc. Rev.* **2012,** *41*, 3340–3380.
36. Diéguez, M., Pizzano, A., Eds. *Metal-catalyzed Asymmetric Hydrogenation. Evolution and Prospect in Advances in Catalysis.* Elsevier: Oxford, 2021; pp. 68.
37. De Vries, J. G., Elsevier, C. J., Eds. *Handbook of Homogeneous Hydrogenation.* Wiley-VCH: Weinheim, 2007.
38. Etayo, P.; Vidal-Ferran, A. Rhodium-catalysed Asymmetric Hydrogenation as a Valuable Synthetic Tool for the Preparation of Chiral Drugs. *Chem. Soc. Rev.* **2013,** *42*, 728–754.
39. Kleman, P.; Pizzano, A. Rh Catalyzed Asymmetric Olefin Hydrogenation: Enamides, Enol Esters and Beyond. *Tetrahedron Lett.* **2015,** *56*, 6944–6963.
40. Tang, W.; Zhang, X. New Chiral Phosphorus Ligands for Enantioselective Hydrogenation. *Chem. Rev.* **2003,** *103*, 3029–3070.
41. Cui, X.; Burgess, K. Catalytic Homogeneous Asymmetric Hydrogenations of Largely Unfunctionalized Alkenes. *Chem. Rev.* **2005,** *105*, 3272–3296.
42. Källström, K.; Munslow, I.; Andersson, P. G. Ir-Catalysed Asymmetric Hydrogenation: Ligands, Substrates and Mechanism. *Chem. Eur. J.* **2006,** *12*, 3194–3200.
43. Roseblade, S. J.; Pfaltz, A. Iridium-catalyzed Asymmetric Hydrogenation of Olefins. *Acc. Chem. Res.* **2007,** *40*, 1402–1411.
44. Church, T. L.; Andersson, P. G. Iridium Catalysts for the Asymmetric Hydrogenation of Olefins with Nontraditional Functional Substituents. *Coord. Chem. Rev.* **2008,** *252*, 513–531.
45. Pàmies, O.; Andersson, P. G.; Diéguez, M. Asymmetric Hydrogenation of Minimally Functionalised Terminal Olefins: An Alternative Sustainable and Direct Strategy for Preparing Enantioenriched Hydrocarbons. *Chem. Eur. J.* **2010,** *16*, 14232–14240.

46. Woodmansee, D. H.; Pfaltz, A. Asymmetric Hydrogenation of Alkenes Lacking Coordinating Groups. *Chem. Commun.* **2011,** *47*, 7912–7916.
47. Zhu, Y.; Burgess, K. Filling Gaps in Asymmetric Hydrogenation Methods for Acyclic Stereocontrol: Application to Chirons for Polyketide-Derived Natural Products. *Acc. Chem. Res.* **2012,** *45*, 1623–1636.
48. Verendel, J. J.; Pàmies, O.; Diéguez, M.; Andersson, P. G. Asymmetric Hydrogenation of Olefins Using Chiral Crabtree-type Catalysts: Scope and Limitations. *Chem. Rev.* **2014,** *114*, 2130–2169.
49. Margarita, C.; Andersson, P. G. Evolution and Prospects of the Asymmetric Hydrogenation of Unfunctionalized Olefins. *J. Am. Chem. Soc.* **2017,** *139*, 1346–1356.
50. Pàmies, O.; Zheng, J.; Faiges, J.; Andersson, P. G. Asymmetric Hydrogenation of Unfunctionalized Olefins or with Poorly Coordinative Groups. *Adv. Catal.* **2021,** *68*, 135–203.
51. Vazquez-Serrano, L. D.; Owens, B. T.; Buriak, J. M. Catalytic Olefinhydrogenation Using N-heterocyclic Carbene–phosphine Complexes of Iridium. *Chem. Commun.* **2002**, 2518–2519.
52. Roseblade, S. J.; Pfaltz, A. Recent Advances in Iridium-catalysed Asymmetric Hydrogenation of Unfunctionalised Olefins. *C. R. Chim.* **2007,** *10*, 178–187.
53. Brandt, P.; Hedberg, C.; Andersson, P. G. New Mechanistic Insights into the Iridium–Phosphanooxazoline-Catalyzed Hydrogenation of Unfunctionalized Olefins: A DFT and Kinetic Study. *Chem. Eur. J.* **2003,** *9*, 339–347.
54. Fan, Y.; Cui, X.; Burgess, K.; Hall, M. B. Electronic Effects Steer the Mechanism of Asymmetric Hydrogenations of Unfunctionalized Aryl-Substituted Alkenes. *J. Am. Chem. Soc.* **2004,** *126*, 16688–16689.
55. Cui, X.; Fan, Y.; Hall, M. B.; Burgess, K. Mechanistic Insights into Iridium-Catalyzed Asymmetric Hydrogenation of Dienes. *Chem. Eur. J.* **2005,** *11*, 6859–6868.
56. Church, T. L.; Rasmussen, T.; Andersson, P. G. Enantioselectivity in the Iridium-Catalyzed Hydrogenation of Unfunctionalized Olefins. *Organometallics* **2010,** *29*, 6769–6781.
57. Hopmann, K. H.; Bayer, A. On the Mechanism of Iridium-Catalyzed Asymmetric Hydrogenation of Imines and Alkenes: A Theoretical Study. *Organometallics* **2011,** *30*, 2483–2497.
58. Mazuela, J.; Norrby, P.-O.; Andersson, P. G.; Pàmies, O.; Diéguez, M. Pyranoside Phosphite–Oxazoline Ligands for the Highly Versatile and Enantioselective Ir-Catalyzed Hydrogenation of Minimally Functionalized Olefins. A Combined Theoretical and Experimental Study. *J. Am. Chem. Soc.* **2011,** *133*, 13634–13645.
59. Gruber, S.; Pfaltz, A. Asymmetric Hydrogenation with Iridium C,N and N,P Ligand Complexes: Characterization of Dihydride Intermediates with a Coordinated Alkene. *Angew. Chem. Int. Ed.* **2014,** *53*, 1896–1900.
60. Helmchen, G.; Pfaltz, A. Phosphinooxazolines—A New Class of Versatile, Modular P,N-Ligands for Asymmetric Catalysis. *Acc. Chem. Res.* **2000,** *33*, 336–345.
61. Metal, T. In *Catalyzed Enantioselective Allylic Substitution in Organic Synthesis;* Kazmaier, U., Ed.; Springer-Verlag: Heidelberg, Berlin, 2012.
62. Butt, N.; Yang, G.; Zhang, W. Allylic Alkylations with Enamine Nucleophiles. *Chem. Rec.* **2016,** *16*, 2687–2696.
63. Grange, R. L.; Clizbe, E. A.; Evans, P. A. Recent Developments in Asymmetric Allylic Amination Reactions. *Synthesis* **2016,** *48*, 2911–2968.
64. Butt, N. A.; Zhang, W. Transition Metal-catalyzed Allylic Substitution Reactions with Unactivated Allylic Substrates. *Chem. Soc. Rev.* **2015,** *44*, 7929–7967.
65. Trost, B. M. Pd and Mo Catalyzed Asymmetric Allylic Alkylation. *Org. Process Res. Dev.* **2012,** *16*, 185–194.

66. Trost, B. M.; Zhang, T.; Sieber, J. D. Catalytic Asymmetric Allylic Alkylation Employing Heteroatom Nucleophiles: A Powerful Method for C–X Bond Formation. *Chem. Sci.* **2010,** *1*, 427–440.
67. Lu, Z.; Ma, S. Metal-Catalyzed Enantioselective Allylation in Asymmetric Synthesis. *Angew. Chem. Int. Ed.* **2008,** *47*, 258–297.
68. Trost, B. M.; Crawley, M. L. Asymmetric Transition-metal-catalyzed Allylic Alkylations: Applications in Total Synthesis. *Chem. Rev.* **2003,** *103*, 2921–2944.
69. Johannsen, M.; Jørgensen, K. A. Allylic Amination. *Chem. Rev.* **1998,** *98*, 1689–1708.
70. Trost, B. M.; Vranken, Van; Asymmetric, D. L. Transition Metal-Catalyzed Allylic Alkylations. *Chem. Rev.* **1996,** *96*, 395–422.
71. Tsuji, J., Ed.; *Palladium Reagents and Catalysis, Innovations in Organic Synthesis.* Wiley: New York, 1995.
72. Pàmies, O.; Margalef, J.; Cañellas, S.; James, J.; Judge, E.; Guiry, P. J.; Moberg, C.; Bäckvall, J.-E.; Pfaltz, A.; Pericàs, M. A.; Diéguez, M. Recent Advances in Enantioselective Pd-Catalyzed Allylic Substitution: From Design to Applications. *Chem. Rev.* **2021,** *121*, 4373–4505.
73. Butts, C. P.; Filali, E.; Lloyd-Jones, G. C.; Norrby, P.-O.; Sale, D. A.; Schramm, Y. Structure-Based Rationale for Selectivity in the Asymmetric Allylic Alkylation of Cycloalkenyl Esters Employing the Trost 'Standard Ligand' (TSL): Isolation, Analysis and Alkylation of the Monomeric form of the Cationic η^3-Cyclohexenyl Complex $[(\eta^3\text{-c-}C_6H_9)Pd\text{- }(TSL)]^+$. *J. Am. Chem. Soc.* **2009,** *131*, 9945–9957.
74. Fristrup, P.; Ahlquist, M.; Tanner, D.; Norrby, P.-O. On the Nature of the Intermediates and the Role of Chloride Ions in Pd-Catalyzed Allylic Alkylations: Added Insight from Density Functional Theory. *J. Phys. Chem. A* **2008,** *112*, 12862–12867.

About the authors

Maria Biosca received her Ph.D. in 2018 at University Rovira i Virgili (URV) under the supervision of Profs. M. Diéguez and O. Pàmies. During her Ph.D., she did a short exchange in the group of Prof. M. Alcarazo (Göttingen University). In 2019, she joined Prof. F. Himo and Prof. K. J. Szabó's groups at Stockholm University as a postdoctoral re- searcher. In 2022, she came back to URV as a Juan de la Cierva postdoctoral fellow, to work in the groups of Profs. M. Diéguez and J. M. Poblet. Her research interests include asym- metric catalysis, water oxidation and DFT- guided catalyst design.

Dr. Maria Besora received her PhD on Computational and Theoretical Chemistry from the Universitat Autònoma de Barcelona under the supervision of Prof. Agustí Lledós and Prof. Feliu Maseras, where she mainly studied the formation of dihydrogen complexes. She did two research stays at the Université de Montpellier II (in Prof. Odile Eisenstein's group) and at University of York (in Prof. John McGrady's group). After a period of post-doc at the University of Bristol (UK) with Prof. Jeremy Harvey studying spin-forbidden reactions, she received a Juan de la Cierva grant and moved to the Institute of Chemical Research of Catalonia (ICIQ) to work at the group of Prof. Feliu Maseras. At the end of 2018, she moved to the Universitat Rovira i Virgili (Tarragona), where she currently holds a position as a Lecturer.

Feliu Maseras obtained his doctoral degree in Chemistry at the Universitat Autònoma de Barcelona (UAB) in 1991. He had a two-year postdoctoral stay sponsored by a European Union fellowship with Keiji Morokuma at the Institute for Molecular Science (Japan), and aftwerwards he worked two years as Research Associate (temporary chargé de recherche) with Odile Eisenstein in Montpellier (France). He has also made several stays as Emerson Center Visiting Fellowat the Emory University (Atlanta, United States). He obtained a tenured position as Associate Professor at the UAB in 1998, where he worked with Agustí Lledós. He has ocuppied this position at UAB until his moving to ICIQ in 2004 as group leader. Author of nearly two hundred publications in scientific journals, his articles have been cited more than 2,000 times in the last 10 years. He received in 2000 the Distinction for the Promotion of University Research (junior category) granted by the Catalan regional government. He got an RSEQ Award in 2011 for his work in the field of Physical Chemistry. His research has focused in the design and application of the quantum mechanics/molecular mechanics (QM/MM) methods to problems of practical interest, and the computational study of molecular systems containing transition metal atoms.

Prof. Oscar Pàmies obtained his Ph.D. in Prof. Carmen Claver's group in 1999 at the Rovira i Virgili University. After three years of postdoc- toral work in the group of Prof. J.-E. Bäckvall at Stockholm University, he returned to Tarragona in 2002. He is currently Professor of Inorganic Chemistry at the Rovira i Virgili University. He received the Grant for Research Intensification from URV in 2008. He has been awarded the ICREA Academia Prize 2010 from the Catalan Institution for Research and Advanced Studies. His research interests are asymmetric catalysis, water oxidation, enzyme catalysis, organometallic chemistry, and com- binatorial synthesis.

Prof. Monterrat Diéguez got her Ph.D. in 1997 at the Rovira i Virgili University (URV). She was post-doc at Yale University with Prof. R.H. Crabtree. Since 2011 she is full professor in Inorganic Chemistry (URV). She is the chair of InnCat research group at URV, succeeding the former chair, Prof. Claver. She is author of 160 articles and 18 books/book chapters with an H index of 46 (2023). She got the Distinction from the Generalitat de Catalunya in 2004 and in 2008 from the URV. She got the ICREA Academia Prize in 2009–14 and 2015–20. Her research interests are catalysis, combinatorial synthesis, artificial metalloenzymes and catalytic conversions of renewable feedstocks.

CHAPTER THREE

Molecular modelling of encapsulation and reactivity within metal-organic cages (MOCs)

Mercè Alemany-Chavarria, Gantulga Norjmaa, Giuseppe Sciortino, and Gregori Ujaque*

Departament de Química and Centro de Innovación en Química Avanzada (ORFEO-CINQA), Universitat Autònoma de Barcelona, Cerdanyola del Vallès, Catalonia, Spain

*Corresponding author. e-mail address: Gregori.Ujaque@uab.cat

Contents

Abstract

Supramolecular chemistry focuses on forming molecular interactions beyond traditional covalent bonds. This discipline provides tools for controlling molecular interactions, having significant impact in various scientific fields, including drug delivery, sensing and catalysis. Supramolecular catalysis is a key area where molecular recognition and encapsulation in host-guest systems can enhance chemical reactions, trying to emulate enzyme efficiency and selectivity. Metal-organic cages (MOCs) are particularly interesting and moldable structures with the ability to encapsulate small molecules and catalyze reactions. This chapter aims to provide a theoretical perspective of encapsulation and reactivity within MOCs by selecting processes that have been studied computationally. The first section focuses on studies detailing the

Advances in Catalysis, Volume 75
ISSN 0360-0564, https://doi.org/10.1016/bs.acat.2024.08.001

molecular mechanisms of the binding process, whereas the second section presents computational examples of reactions that are accelerated by MOCs. In the final section, general conclusions and discussion on potential future directions in this field are presented.

1. Introduction

Supramolecular chemistry is a well-established discipline mainly based on establishing networks of molecular interactions. Formation of these molecular interactions goes beyond the scope of traditional covalent bonds since non-covalent forces play a crucial role to organise the formation of dynamic and functional molecular assemblies *(1)*. Supramolecular chemistry offers a versatile toolbox for scientists to manipulate and control molecular interactions, paving the way for innovative solutions in various scientific disciplines *(1b)*. This field covers the study of host-guest chemistry, molecular recognition, self-assembly, and the design of functional materials with applications ranging from drug delivery and sensing to catalysis and nanotechnology.

Among its many applications, supramolecular catalysis emerges as a fundamental topic, where molecular recognition and encapsulation drive catalytic processes within tailored host-guest systems *(2)*. Supramolecular catalysis, as defined by Hosseini and Lehn, involves the chemical transformation of a bound substrate, relying on complexation and recognition steps as prerequisites *(3)*. It operates on the principles of molecular recognition, resembling the active sites of proteins, with the ability of guiding and accelerating chemical reactions. The supramolecular catalysts, often comprised of host molecules and guest substrates, form dynamic and reversible complexes driven by non-covalent interactions (NCIs) such as hydrogen bonding, electrostatic and hydrophobic interactions, etc. *(4)*. For guests acting as reactants they can exhibit very high catalytic efficiency and selectivity, in some cases comparable to those of enzymes *(5)*. The local microenvironment within the cavity can differ substantially with the surrounding bulk in solution thus modifying the reaction rates *(6)*, similarly to enzymes *(7)*.

Over the past decades, the design and synthesis of synthetic hosts have significantly expanded the repertoire of available supramolecular entities. Among the most fascinating developments are metal-organic cages (MOCs), also known as metallocages, supramolecular organometallic complexes, porous organic cages, coordination cages, etc. They are formed through the self-assembly of metal ions or clusters with organic ligands, primarily driven by coordination principles *(8)*. Metallocages are discrete individual molecules featuring a cavity within their structure. Notably, they exhibit diverse topologies and define nanoscale architectures, offering attractive opportunities for selective and specific hosting of small molecules in their cavities *(5c,9)*. Tailoring the cavities of MOCs has led to a plethora of applications, including molecular recognition, product mixture purification, gas sorption, drug and photosensitizer delivery, and catalysis *(2a,9d,10)*. We are interested here in coordination cages that can act as molecular flasks, employing their cavities to regulate chemical reactions for encapsulated reactants *(11)*. The number of catalytic processes facilitated by MOCs are constantly increasing. In this chapter, among those that have been computationally investigated, we decided to present a selection aiming to show how theoretical methods lead in our molecular understanding of these processes.

For elucidating catalytic mechanisms at the molecular scale, providing roads to explore bond formation, bond cleavage, and host-guest interactions computational techniques are indispensable *(12)*. Despite their significance, theoretical studies on the rate accelerated reactions within supramolecular hosts remain relatively limited, although they are lately increasing *(13,14)*. Regarding molecular design of metallocages, advanced computational tools have emerged to predict their shapes and molecular properties, but they are not the purpose of this chapter *(15)*.

This chapter aims to provide an overview of chemical processes catalysed by MOCs from a computational perspective. The first section is dedicated to describing studies that provide a molecular description of the binding process (Fig. 1A). In the second section, selected examples of reactions accelerated in the presence of MOCs that have been computational studied are described (Fig. 1B). Finally, we present general conclusions and provide some insights into future directions on the field.

2. Encapsulating the reactants—Host-Guest binding

Molecular recognition refers to the specific and selective binding interaction between molecules and is mainly based on complementary structural and chemical features *(16)*. In this process, molecules (hosts and guests) recognize each other through NCIs such as hydrogen bonding, van der Waals forces, electrostatic interactions, etc. This recognition leads to

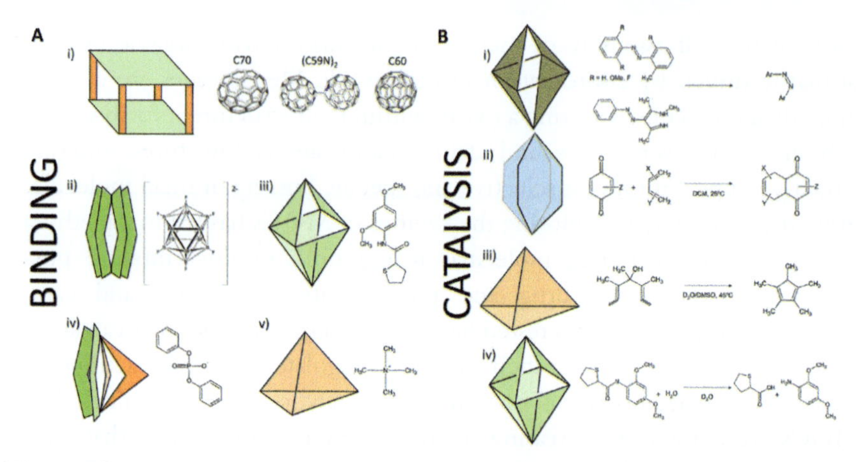

Fig. 1 Schematic representation of the selected MOCs covered in this chapter: (A) MOCs implicated in binding studies, including samples of guests, and (B) MOCs along with a selected sample of accelerated chemical reactions.

the formation of stable complexes with well-defined geometries and binding affinities *(17)*. Understanding the binding (encapsulation process) between host and guest molecules is fundamental and computational investigation has emerged as an indispensable tool for this analysis.

Despite the crucial relevance of molecular interactions, the accurate computational evaluation of the binding involving two molecular species in a biological and/or chemical context remains a significant challenge *(18)*. For typical hosts (i.e. cyclodextrins, cucurbiturils, etc.) they have been largely studied *(19)*. This section outlines the computational methodologies employed to calculate the host-guest binding energies, mainly focussed in those cases where hosts employed are metallocages; then, selected examples are exposed to illustrate the results.

2.1 Computational methods for simulating Host-Guest binding

There are several computational methods available to calculate binding energies that are commonly employed for proteins and for other host-guest complex *(18,20)*. Each of these methods has its own advantages, limitations, and applicability and the selection generally depends on the accuracy needed, the time required (in computational and human terms), and the computational resources available.

Calculating the binding Gibbs energies ($\Delta G_{binding}$) of flexible host-guest systems in gas phase would be affordable by determining the minimum energy of all the moieties involved: the host (reactants), the guest (metallocage), and the host-guest (reactant-metallocage) complex, or even for small hosts in solution. Post-Hartree-Fock or density functional theory (DFT) based methods could be employed to determine these energies in gas phase *(21)*. In solution or for larger hosts, however, it requires averaging over all accessible solvent-solute configurations. Computational techniques based on statistical methods (molecular dynamics, MD) employing empirical force-fields to calculate the energies are the most employed ones. This is mainly because they allow large simulation times as well as wide exploration of the conformational and configurational space of the system *(22)*. Note that Monte Carlo simulations are an alternative method but none of the samples here commented uses this technique, so it will not be further commented *(23)*.

The most commonly employed methods for computing binding energies are the following:

1. *Molecular docking* involves predicting the preferred orientation and binding mode of a molecule (ligand or guest) within a receptor (protein or host) binding site *(24)*. Docking is based on the exploration of the conformational space of the ligand and receptor, the latter usually with limited flexibility. Docking algorithms search for energetically favourable configurations by employing approximate scoring functions based on intermolecular interactions and complementarity *(25)*.
2. Related approaches include *linear interaction energy (26)*, *molecular mechanics/generalized Born surface area (MM/GBSA)(27)* as well as *molecular mechanics/Poisson-Boltzmann surface area (MM/PBSA) (27b,28)* methods. These are end-states methods that estimate binding energies based only on bound and unbound states. They decompose the total energy into contributions from various components, including van der Waals, electrostatic, and solvation energies.
3. *Molecular Dynamics* (MD) simulations compute the trajectories of atoms and molecules over time, allowing for the exploration of the dynamic behaviour of protein-ligand or host-guest complexes. MD simulations can estimate binding energies by analysing the interactions between the ligand and receptor over a simulation trajectory. The most accurate methods include Metadynamics *(29)*, Umbrella sampling (US) *(30)* and Free Energy Perturbation (FEP) *(31)*. These methods can reach high accuracy, but their computational cost is significantly higher than in the previous methods.
4. Machine Learning and Artificial Intelligence-based approaches have led to the development of predictive models for estimating binding energies

based on large datasets of protein-ligand *(32)*. These methods use molecular descriptors, structural features, and experimental data to train models capable of accurately predicting binding affinities.

Each computational method has its strengths and weaknesses, and the choice of method depends on factors such as the system's size, complexity, and the level of accuracy required. The methods employed to obtain accurate binding energies, and in some cases the energy profiles for the encapsulation process, are mostly based on MD simulations. Nevertheless, they are not conventional MD simulations because binding events typically occur on timescales ranging from milliseconds to seconds, and conventional MD alone hardly ever reproduce these events. Therefore, rare event sampling methods such as US, accelerated MD (aMD), and metadynamics are generally employed. Contrary to the FEP approach, which connects the end states through an alchemical pathway, these techniques enable the modelling of the host-guest encapsulation process, providing access to both thermodynamics and kinetics parameters.

US is a computational method utilized to determine the Gibbs energy profiles along a specific reaction coordinate (such as the distance between two molecules in a binding process) *(30)*. In this method, the reaction coordinate is divided into small intervals, and simulations are performed at each interval using biasing potentials. These potentials act to restrain the system along the reaction coordinate, allowing for thorough sampling of the entire range of configurations. By performing multiple simulations with different potentials, US generates a series of overlapping histograms of the reaction coordinate, which are then combined to obtain the unbiased Gibbs energy profile; for that, statistical reweighting techniques such as the Weighted Histogram Analysis Method are employed *(33)*. This method provides accurate insights into the ligand binding process.

aMD is a computational technique used to enhance the sampling of rare events and overcome energy barriers in MD simulations *(34)*. In aMD, an additional boost potential is applied to the potential energy surface, which accelerates the system's dynamics by lowering energy barriers and facilitating transitions between different states. By incorporating the boost potential into the simulation, aMD enables the exploration of the full conformational space of the system, including energetically unfavourable states that are typically inaccessible in conventional MD simulations. This enhanced sampling allows for accurate estimation of binding energies.

Metadynamics is also a powerful computational technique to explore complex Gibbs energy landscapes in molecular systems, including the calculation of binding energies *(29)*. In metadynamics, an additional bias potential is applied to the system, which depends on a set of collective variables (CVs) that describe the relevant degrees of freedom of the system. This bias potential is deposited along the chosen CVs being dynamically updated to discourage the system from revisiting previously explored regions of the CV space. This encourages the system to escape from energy minima and explore different states, facilitating the sampling of rare events and transitions. By analysing the accumulated bias potential, one can reconstruct the underlying Gibbs energy landscape and extract associated thermodynamic information, such as binding energies or obtaining the Gibbs energy profile for the process. Metadynamics offers several advantages, including its ability to efficiently sample complex energy landscapes and its versatility in studying a wide range of molecular processes. However, it requires careful selection of CVs and tuning of parameters to ensure accurate results. Overall, it is a powerful tool for investigating molecular recognition and binding phenomena in biological and chemical systems.

2.2 Selected examples from literature

In this section are gathered selected examples from literature where the binding (or unbinding) process of a ligand into a guest receptor is analysed by means of computational methods. For each example, the binding process as well as the methods employed will be described.

The group of Schäfer has been among the most active ones in analysing the encapsulation process in several MOCs *(35)*. For instance, they investigated the binding process of the $[B_{12}F_{12}]^{2-}$ guest into $[Pd_2L_4]^{4+}$-type metallocages. This MOC, designed at the Clever's group, consists of 2Pd (II) ions with 4 photochromic dithienylethene (DTE) banana shaped ligands (L) coordinating in a square planar configuration. DTE has two possible photoisomers, a blue coloured closed isomer (C) and a colourless opened isomer (O); see Fig. 1 *(36)*. The encapsulation of the guest $[B_{12}F_{12}]^{2-}$ into this positively charged host was computationally investigated in explicit acetonitrile solvent using US based on a force field. They were able to obtain the Potential Mean Force (PMF) of the process along the reaction coordinate (here being the z-component of the vector defined from the host's centre of mass to the guest's centre of mass), limiting the motion of unbound guest to a plane orthogonal to the reaction coordinate by the use of a restraint (Fig. 1C) *(35)*.

The PMF to evaluate the encapsulation process and calculate the binding Gibbs energy was investigated for each of the possible cage configuration (depending on the photoswitching events), having a special consideration of the releasing routes the guest can follow in an intermediate combination cage. It is not the same to exit the cage between O–O DTE ligands, C–C, or O–C ones (Fig. 1B). All these configurations encapsulate spontaneously the guest, being 4 O the one with the best affinity and 4 C the one with the least affinity, although other intermediate conformations, as 3O1C, 1O3C and 2O2C (*trans* and *cis*) combinations are also possible. They observed, in agreement with experimental results *(37)*, that the cyclization of the first ligand (transitioning from 4 O to 3O1C configurations) represents the biggest affinity decrease and it is probably the situation where the guest is ejected from the metallocage.

By means of experimental NMR and isothermal titration calorimetry (ITC) observations the authors concluded that the binding process is entropy driven *(38)*. Schäfer and co-workers were able to separate the binding Gibbs energy into binding entropy and binding enthalpy observing that the binding enthalpy was positive (thus it is not an enthalpy favourable process), and hence the process is entropy driven, in agreement with ITC data. The enthalpy was further analysed in terms of van der Waals, electrostatic, and bonded interactions, observing that the last ones are the main contributors into the positive enthalpy value of the encapsulation process. The reasoning they put forward is that the host-guest complementarity is probably not good enough and it is causing steric tension to the cage in the bound state. They also noted that the electrostatic enthalpic factor was not contributing as expected in making the bound state favourable. Regarding entropic effects, they stated that the negative entropy that makes the binding process spontaneously possible is likely linked to the changes in the guest's solvation that occur during the binding.

In a subsequent work by the same group they extended their investigation of this system, this time employing DFT calculations *(39)*. They studied how the photoswitchable nature of the DTE ligand influences the binding of $[B_{12}F_{12}]^{2-}$ and two other guests of different size and electrostatic distribution, and how they affect the metallocage properties and dynamics.

The group of Clever also developed a class of heteroleptic metallocages combining different type of ligands on the same metallocage. Concretely, they worked on a lantern-shaped metallocage, consisting of 2Pd(II) ions coordinating in a square-planar fashion with 4 ligands, 2 Ligands A that can be highly functionalized and can act as hydrogen donors, and 2 Ligands B with aromatic character (Fig. 2) *(40)*; these cages [$Pd_2A_2B_2$], will be labelled as cage-NH from now on. Self-assembling systems usually contain only one kind of organic molecule acting as ligands, thus, heteroleptic metallocages are not commonly designed. Cage-NH has been experimentally characterized and its binding affinity has been determined through NMR for several phosphate diesters guests (40a).

In order to determine if the hydrogen bonding aspect is the driving force for guest encapsulation they also tested a Ligand A with no capability of establishing this kind of interactions substituting H by Me (cage-Me). Schäfer and co-workers were able to calculate the $\Delta G_{binding}$ by means of performing long MD simulations (up to 50 μs in explicit DMSO solvent), where they observed that the guest is encapsulated and released several times. Counting the binding and unbinding events and applying the proper statistics *(41)*, they could estimate the binding Gibbs energy, with values no higher than −3.3 kcal/mol, in agreement with experimental observations.

Among these heteroleptic metallocages, those with ligands able to make hydrogens bonds showed more affinity than the methylated ligand, although the binding affinity differences are surprisingly small (in the range of 0–1.5 kcal/mol). An analysis of the contacts taking place during the MD trajectories when the guest is encapsulated was employed to evaluate the weight of non-covalent interaction in the process. Their conclusion was

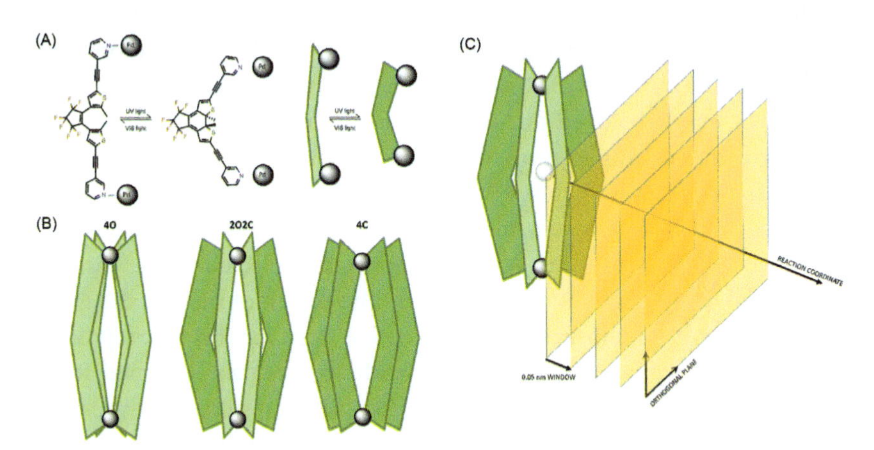

Fig. 2 (A) Schematic representation of opened (O, in light green) and closed (C, in dark green) conformations of the lantern-shaped $[Pd_2L_4]^{4+}$ cage with coordinating DTE ligand. (B) Three possible metallocage structures with different proportion of O and C ligands are showed (4O, 2C2O, 4C). (C) Representation of Umbrella Sampling's reaction coordinate and the orthogonal planes defined by restrain to limit the guests movement in unbound state.

that, although hydrogen bonds are important for the encapsulation, they can be partially compensated through hydrophobic interactions, thus allowing the encapsulation to take place in all cases. The same group also modelled a mixed cage including a ligand able to establish hydrogen bonds and a methylated ligand *(40b)*. This was considered to be and intermediate structure between cage-NH and cage-Me, referenced as cage-H, and had not been experimentally characterized.

The calculated $\Delta G_{binding}$ show that cage-H's affinity is lower than cage-NH but higher than cage-Me. With the acquired knowledge they proceeded to apply Markov State Model (MSM) to the MD trajectories to determine the dynamics and possible pathways for encapsulation of the guest into the three different cages. MSM is a master mathematical framework that has been used to model and understand the dynamics of molecular systems over extended timescales *(42)*, but not previously employed in supramolecular systems *(40b)*. Schäfer defined 6 states in total for each cage, 4 bound states, 1 transitioning from inside to outside, and 1 completely unbound state. The analysis provided information on which states were more prevalent and which transitions between states were more favourable, thus getting a detailed picture of the dynamics of each system and being able to extract conclusions on how the cage structure and features influence the binding affinity for a specific guest.

The binding of several cationic guests inside the supramolecular tetrahedron metallocage, $[Ga_4L_6]^{12-}$, using classical MD simulations was investigated by our research group *(43)*. The attach-pull-release (APR) method *(44)* was employed to calculate the absolute binding Gibbs energy for the host-guest complexation in explicit solvent. The effects of the non-standard force field parameters of the host on the calculated binding Gibbs energy were evaluated systematically. The US technique *(45)* was used to calculate the PMF along the reaction coordinate selected as the distance between the centre of mass of the host and the centre of mass of the guest. The calculated PMF showed two consecutive steps for the entire process: (i) exterior association of the guest to the host (ion-pair intermediate or external binding of the guest by the host) and (ii) encapsulation of the guest (internal binding of the guest by the host). In the transition zone for the binding of the guest (NEt_4^+) into the metallocage we identified the presence of K^+ ions, suggesting that there is a replacement of ions during encapsulation, although it could not be proved by our simulations. The calculated Gibbs energy profile was in very good agreement with experimental measured ones *(46)*.

Ribas, Feixas, Jiménez-Barbero and co-workers *(47)* used MD *(48)* and aMD simulations *(34,49)*, in combination with experimental techniques, as ^{1}H–^{1}H exchange spectroscopy (2D-EXSY) NMR, to study how a positively charged and very flexible Pd(II)-based nanocapsule with four entry gates can recognise and encapsulate fullerenes (Figs. 3A and 3B). To elucidate the essential details of the recognition process at the molecular level, MD simulations of the nanocapsule in the presence of fullerenes were performed (30 replicas accumulating a total of 75 μs of simulation time). With this vast quantity of molecular data, they were able to obtain several complete encapsulation events; then, they analysed the structural behaviour of the cyclic groups present at the entries that act as a gate. In the transition of the unbound to the bound complex they identified several intermediate states. They identified the placement of the guest into the host, its rotational movement inside its cavity, as well as deciphering several binding pathways. They also observed that after complete encapsulation the nanocapsule becomes more rigid, stabilizing the bound state and slowing down the unbinding process. Moreover, using aMD they were also able to assess competitive encapsulation

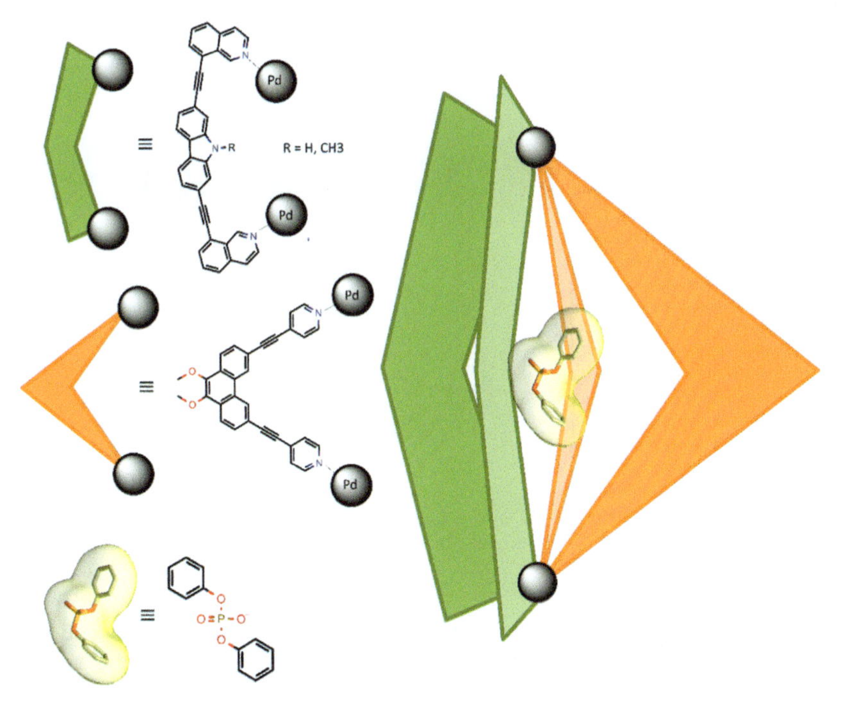

Fig. 3 Diagram of the heteroleptic metallocage $[Pd_2A_2B_2]$ with the organophosphate diphenylphosphate encapsulated. Ligand A is depicted in green, ligand B in orange.

among C60 and C70. The authors showed that the combination of NMR, MD, and aMD methodologies provides a precise understanding of the subtle events that govern encapsulation, making it a predictive tool for analysing host–guest interactions in various supramolecular systems.

Fujita, Takezawa and co-workers showed that the reaction rate for the hydrolysis of amides was significantly increased within octahedral $[Pd_6L_4]^{12+}$ metallocages (Fig. 4) *(50)*. The computational study by means of a multiscale simulation approach performed by the Pavan's group significantly helped to understand the process from a molecular point of view, as explained below *(51)*. In this section though, we will only comment on the encapsulation analysis they performed.

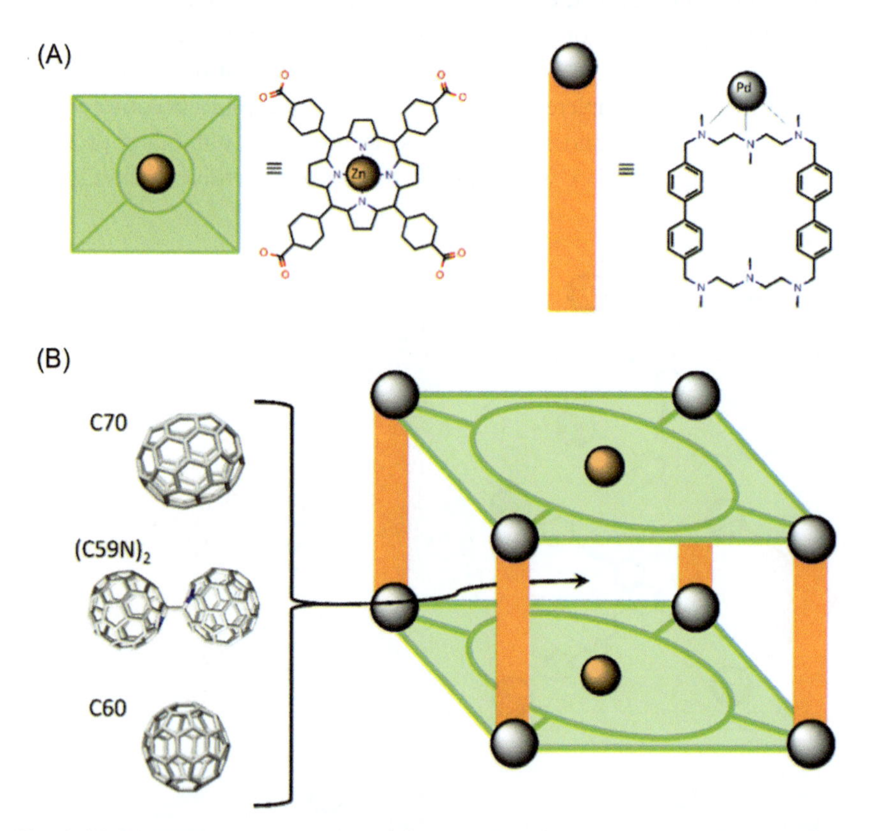

Fig. 4 (A) Schematic representation of the nanocapsule's components. Roof and floor consist of Zn coordinating porphyrins that coordinate with Pd(II) at the vertices (green). Pillars of the cage are depicted in orange and represent flexible structures, also coordinating Pd(II). (B) Picture of the assembled nanocapsule and the fullerene guests studied.

The authors employed Well-Tempered Metadynamics (WT-MTD) and Infrequent WT-MTD (IF-MTD) to analyse the encapsulation process. IF-MTD differs from classical metadynamics on how the bias is applied to the system. In IF-MTD the bias is added to the CV according to previously specified criteria, for this reason deep knowledge of the system in question is needed in order to use this technique. On one hand IF-MTD can bring efficiency and accuracy to the calculations of sporadic events, while on the other hand, if the criteria is not adequate, it can lead to deficient sampling of the Gibbs energy landscape or requiring longer simulation times to reach convergence, among other counterpoints.

In order to understand how the amide in question enters and exits the supramolecular host they defined three CV: (1) Distance between the guest's centre of mass and the host's centre of mass; (2) Mean contacts between atoms of the guest and the atoms of the host; (3) standard deviation of contacts between the guest's atoms and the host's atoms. These three CV were necessary to reach convergence for this system. As the authors explained in the paper, it is a challenging task to accurately describe a system in a low number of CVs so that an MTD simulation can converge properly. They obtained $\Delta G_{binding}$ as well as the Gibbs energy barriers ($\Delta G^{\ddagger}$) associated with the encapsulation/expulsion in the cavity of $[Pd_6L_4]^{12+}$ metallocage of the amide (in their *cis* and *trans* conformations). Moreover, they were able to obtain these values in two different situations: for the system hosting and non-hosting an additional non-reacting guest. The results showed that increasing crowding in the metallocage cavity stabilizes more the amide in the *cis* conformation, the most reactive one. Overall, they obtained computational data that is in good agreement with the experimental binding studies for this system giving a molecular description of the process *(50)*. Their computational approach can also be applied to other systems to study how guest and host modifications can influence reactivity inside the confined space.

In summary, the binding and unbinding processes of hosts in various metallocages using computational methods have been pointed out. The encapsulation of $[B_{12}F_{12}]^{2-}$ in $[Pd_2L_4]^{4+}$ metallocages, was investigated using US to derive the PMF and binding Gibbs energy. They found that the process is entropy-driven, and the binding enthalpy is primarily influenced by van der Waals and bonded interactions. The use of long MD simulations to estimate $\Delta G_{binding}$ was employed to study binding on heteroleptic metallocages, revealing that hydrogen bonding enhances affinity but can be compensated by hydrophobic interactions. Classical MD simulations based on APR method were applied to examine cationic guest binding in $[Ga_4L_6]^{12-}$ metallocages, in very good agreement with experiment. To study fullerene encapsulation in Pd(II)-based nanocapsules, aMD simulations were employed with great success to describe the process and obtain the associated Gibbs energies. These studies show the effectiveness of computational to understand host-guest binding process in supramolecular systems.

3. Reactivity within metallocages

Supramolecular hosts offer synthetic chemists a versatile instrument to direct non-covalent macromolecular reactivity, particularly outside of biological contexts. In this sense, MOCs are molecular entities characterized by an internal cavity within their structure. Remarkably, they show a variety of topologies and precise nanoscale arrangements, presenting appealing options for the selective and precise encapsulation of molecules within their cavities. Within these assemblies, remarkable reactivity has been observed, with catalysts achieving surprisingly high rate accelerations or even facilitating challenging transformations under mild conditions *(6,10d)*.

Unlike conventional catalysts requiring complex ligand scaffolds, supramolecular catalysts self-assemble from simple components, offering tailored microenvironments even under unfavourable conditions. These exceptional properties emphasize their significance in synthesis and warrant further investigation into their applications, where catalysis is one of the most valuable ones. This section of the chapter gathers selected computational studies over the reaction mechanisms taking place within these supramolecular hosts that give rise to exceptional reactivities.

3.1 Computational methods for simulating reactivity in confined space

Studying chemical reactions, which involve the formation and breaking of chemical bonds, necessitates the application of quantum mechanics (QM). When dealing with MOCs, the size of the systems is quite large reaching the limits of QM calculations (nowadays around 300–350 atoms for studying reaction profiles). The insertion of explicit solvent molecules significantly aggravates this limitation. On other side, the potential flexibility and accessibility to multiple configurations (both for hosts and guests) make statistical methods well-suited for analysis. Hence, a combination of methods is typically required rather than relying solely on one approach. In this overview, we summarize the methods or combinations thereof utilized in existing literature.

The widespread use of QM, particularly those based on DFT, has become prevalent *(52)*. QM approaches enable the description of molecular systems at the electronic level, allowing for precise predictions of structure, energy, and molecular properties. They find extensive application in inorganic and organometallic chemistry due to the complexity introduced by metals *(53)*, and have been effectively employed in studying catalytic metallocages *(54–56)*.

Despite the increase in computational power, QM methods remain computationally expensive. Consequently, various strategies have been devised to address catalytic metallocages from a QM perspective:

1. Cluster models: this is a particularly drastic approach that involve truncating a portion of the cage to generate cluster models, thereby circumventing consideration of the entire system's size.
2. the utilization of Hybrid QM/MM methods. This is an alternative of high value, wherein a portion of the system, typically the reactive component, is treated using QM formalism while the remainder is handled at the MM level *(57)*.
3. Semi-Empirical Quantum Mechanical (SQM) methods is an alternative that has found applications due to their significantly faster computational speeds compared to ab initio methods. Recent advancements include modifications to density functional tight binding methods such as GFN-xTB *(58)*, among others *(59)*.
4. Alternatively, Force Fields (FF) approaches, commonly employed in structural studies of organic and biological systems, can be adapted for metallocages by incorporating custom-generated metal parameters into their FF *(60)*, although they are not suitable for studying reactivity. As previously stated, they are mainly employed when working with computational techniques based on statistical methods as MD. They are utilized to address conformational and binding issues, rather than reactivity matters.
5. Machine Learning Potentials (MLPs) has emerged as a powerful tool for rapidly and accurately describing multi-dimensional potential energy surfaces *(61)*, thus containing bond-breaking bond-forming processes *(62)*. MLPs offer computational efficiency and eliminate the need for expensive QM methods, and unlike empirical FFs that require

predefined functions, MLPs can extract complex patterns from large datasets without such constraints. They are potentially capable of describing all types of bonding and atomic interactions, applicable to both reactive and non-reactive systems. Nevertheless, despite their potentiality there are no cases yet applied to MOCs.

Catalytic reactions involving MOCs typically occur in solution, necessitating consideration of the medium's effects in their simulations. The influence of the bulk environment can be modelled using explicit, implicit, or hybrid explicit-implicit solvation approaches. However, incorporating full explicit solvation shells within metallocages significantly increases the number of atoms involved, limiting this approach to SQM or FF based methods, seriously restricting their applicability to study reaction mechanisms, especially for the latter. From a quantum mechanical perspective, hybrid model combining a reduced number of explicit solvent molecules with continuum models achieves a favourable balance between accuracy and computational efficiency has been commonly employed, ab initio MD (AIMD) have been also occasionally applied. Self-Consistent Reaction Field *(63)*, which treats the interaction between solute and solvent as a classical electrostatic problem is one widely adopted implicit approach. Commonly used methods include the polarisable continuum model *(63)*, the solvation model based on solute electron density (SMD) *(64)*, and the conductor-like screening model *(65)*.

3.2 Selected examples from literature

In recent years, computational modelling has emerged as a powerful tool for investigating reactions occurring within MOCs *(14)*. In this section, we expose a comprehensive examination of several prominent examples from the literature where computational methods have been employed to explore and elucidate the mechanisms of reactions occurring within MOCs. Through detailed analysis and discussion of these studies, we aim to highlight the significant contributions of computational modelling in advancing our understanding of the intricate chemistry occurring within these confined spaces. We are grouping the investigated chemical reactions based on the main geometries exhibited by the MOCs utilized to enhance their efficiency. They are gathered in Fig. 5.

3.3 Lantern-shaped cages

In 2018, Duarte, Lusby and co-workers reported an atomistic analysis of the Diels-Alderase activity of two Pd_2L_4 metallocages ($[Pd_2L_4^N]^{4+}$ and

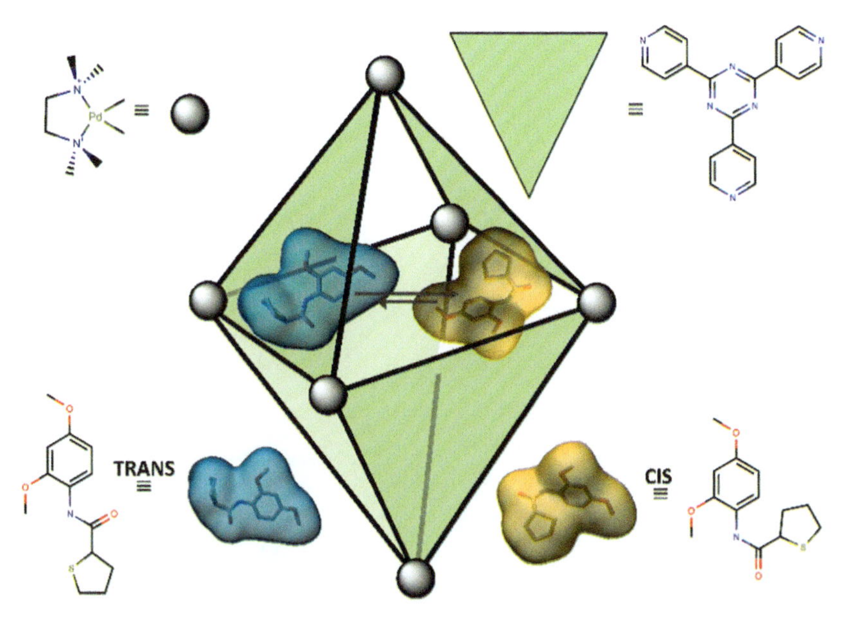

Fig. 5 Scheme of $[L_4Pd_6]^{12+}$ metallocage where four ligands coordinate with 6Pd(II) atoms. Amide studied is shown encapsulated, with both cis and trans isomerization; the cis conformation is the most accessible after binding.

$[Pd_2L_4CH]^{4+}$, with L^N = 2,6-bis(3′-pyridylethynyl)pyridine, and L^{CH} = 1,3-bis(3′-pyridylethynyl)benzene) involving a series of 14 quinones as dienophiles (Fig. 6) *(66)*.

They combined classical MD and DFT outcomes to rationalize the experimental host-guest binding affinities and catalytic performance of the MOCs. The authors also derived a general computational protocol to estimate the Diels-Alderase activity within the metallocage. This approach, tested on a dataset of six quinones and five dienes, achieved an 80% accuracy rate.

Classical MDs in explicit solvent using the cationic dummy atoms approach for Pd(II) *(67)*, outlined a considerable degree of helical distortion of the metallo-cage. The binding energies of anthraquinone and benzoquinone were obtained by static DFT simulations both in vacuum and in solvent (SMD). Including solvent effects was key to matching the highest experimental affinities of both substrates with $[Pd_2L_4^N]^{4+}$. The theory level, benchmarked toward the complete guest's series pointing out the DFT (M06) as the best method (MAD of 2.2 and 1.7 kcal/mol, respectively). Semi-empirical PM7 and tight-binding GFN-xTB methods showed poor

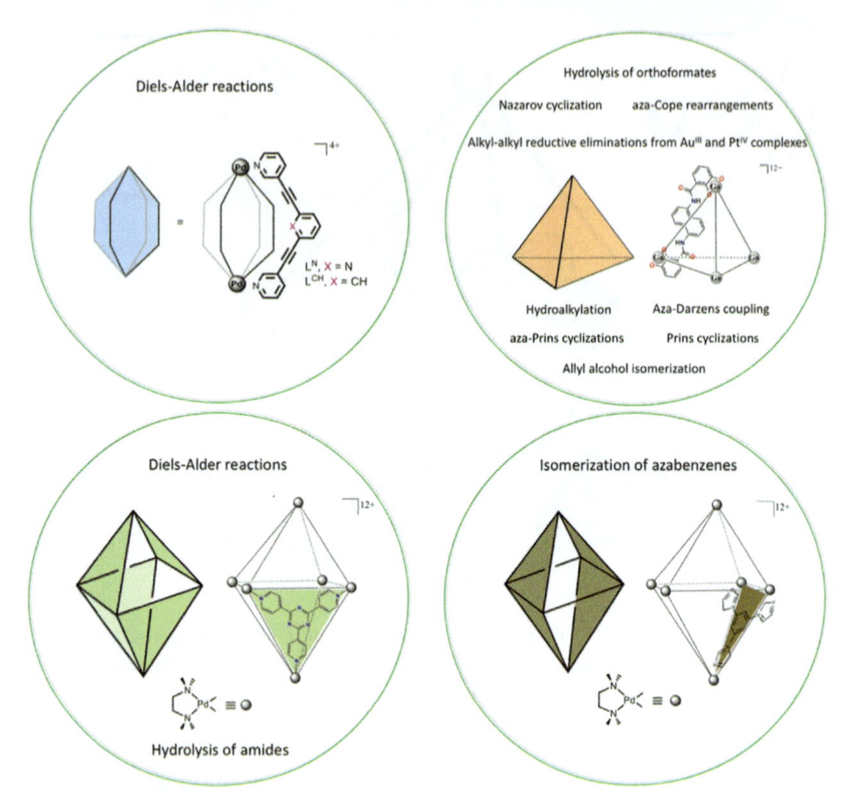

Fig. 6 Schematic representation of the main metallocages able to accelerate reaction rates.

correlation with experimental data. The high affinities of quinones toward $[Pd_2L_4]^{4+}$ were explained by analysing the host-guest NCIs within the metallocage structures, identifying the CH(benzoquinone)$\cdots\pi(\mathbf{L^{CH}})$ (for LCH look at Fig. 6) NCI as the main contribution to the enhanced binding energy of $[Pd_2L_4^N]^{4+}$.

The Diels-Alder reaction between benzoquinone (**bq**) and isoprene (**d1**) inside both homologue metallocages was analysed at DFT theory level. Forming either the inclusion complex **d1**$\subset[Pd_2(L \text{ or } L^N)_4]^{4+}$ or (**d1** + **bq**)$\subset[Pd_2(L \text{ or } L^N)_4]^{4+}$ was not favourable. Instead, the study found an intermolecular reaction mechanism between **bq**$\subset[Pd_2(L^{CH} \text{ or } L^N)_4]^{4+}$ and **d1**. The reaction's activation energy was significantly lower compared to the reaction in bulk only for the case of $[Pd_2(L^N)_4]^{4+}$. The rate acceleration was estimated in 4.7 kcal/mol, obtaining an activation energy barrier of 18.8 kcal/mol consistent with experimental activation energy of 20.4 kcal/mol. The reactivity differences observed in the metallocages were rationalised by energy distortion/interaction decomposition and NCI analyses. Their findings showed significant distortion in the cages at the transition state geometry, with $[Pd_2L_4^N]^{4+}$ requiring less distortion than $[Pd_2(L^{CH})_4]^{4+}$. NCI analysis highlighted larger steric clashes in $[Pd_2(L^{CH})_4]^{4+}$ and favourable CH···N hydrogen bonds in $[Pd_2L_4^N]^{4+}$.

The same authors, in a subsequent study, analysed the different catalytic performance of the $[Pd_2(L^{CH})_4]^{4+}$ and $[Pd_2(L^N)_4]^{4+}$ toward Michael addition reactions (Fig. 7). In contrast to the Diels-Alderase activity, the N-cage was found to be catalytically inactive, whereas the CH-cage was proven to be active. Moreover, the $[Pd_2(L^{CH})_4]^{4+}$ host demonstrated complete diastereo-selectivity *(68)*.

A computational analysis, based on static DFT exploration of the cavity's electrostatic potential (ESP), first addressed the reason behind the inertness of the $[Pd_2(L^N)_4]^{4+}$. Within this host, in contrast to the catalytically active $[Pd_2L_4CH]^{4+}$, the ESP is reduced by the central nitrogen atoms which hampers its ability to stabilize negatively charged intermediates generated by intermolecular attack of non-bound anionic nucleophiles. Similar analysis also provided evidence that the ESP is also responsible for the cage's ability to shift the reduction potential of bound quinones by more than 1 eV, corresponding to a stabilization of the quinone radical anion by 23 kcal/mol *(69)*.

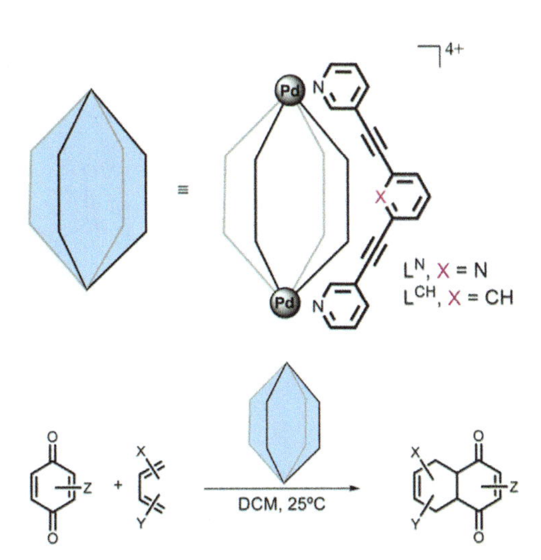

Fig. 7 Representation of the $[Pd_2L_6]^{4+}$ lantern-shaped MOC on the top, and Diels-Alder catalysed reaction by this host on the bottom.

Further DFT-based studies, focused on the stabilization of the epimeric diastereomers (RR/SS) or opposite (RS/RS) of intermediate **I** within the $[Pd_2(L^{CH})_4]^{4+}$ MOC (Fig. 7, bottom). Each diastereomer can take on one of two conformations of the nitroester and methine groups either *syn* or *anti* periplanar. In the absence of the cage, the energy difference between the two conformations of each diastereomer is around 3.5 kcal/mol. However, with encapsulated species, the range of relative energies increases up to 12.8 kcal/mol with an energetic preference by 4.8 kcal/mol for the RR/SS *anti* periplanar diastereomer. This stabilization, by positioning the acidic α-proton of the nitroester-methine on the same face, drives the stereo-selection toward the *anti*-cyclohexyl conformation observed.

3.4 Octahedral-shape cages

Fujita and co-workers pioneered the work on metallocages and their applications to enhance reaction rates *(11b)*. The most outstanding ones are shown in Fig. 8.

The Diels–Alder reaction between 9-hydroxymethylanthracene (**1a**) and N-cyclohexyl maleimide (**1b**) occurs in bulk solution at the 9,10-position of the central ring **1a**. The confined space of the bowl-shaped cage, significantly accelerate the process. However, when using the octahedral cage, the regioselectivity of the reaction is shifted toward the 1,4-position of the terminal ring of **1a** (Fig. 9) *(70)*. Xu and co-workers *(71)*, recently reported a

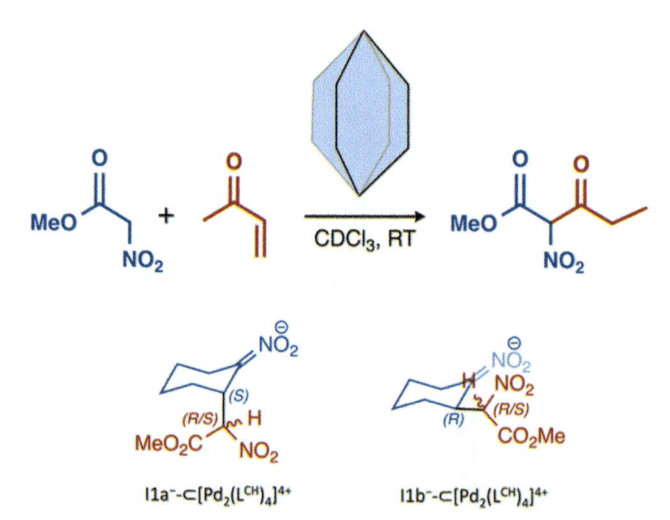

Fig. 8 Representation of the $Pd_2L_2^{4+}$ host lantern-shaped MOC catalysed Michael addition on the top, and epimeric diastereomers of intermediate **I** on the bottom.

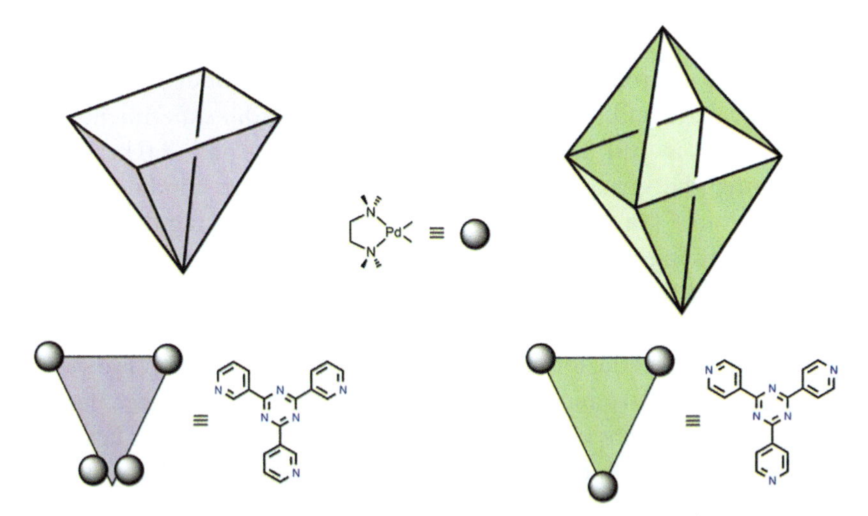

Fig. 9 Representation of the bowl-shaped (left) and octahedral (right) $Pd_4L_4^{12+}$ host MOCs.

QM/MM DFT-based (B3LYP-D3(BJ):PM6D3) analysis of Diels-Alderase activity shown by two related hosts, the bowl-shaped and octahedral topoisomers of $[Pd_6L_4]^{12+}$, reported by Fujita (Fig. 8).

The Diels–Alder reaction pathway in bulk solution involves the approach of substrates **1a** and **1b**, which is slightly endergonic, leading to product **2** driven by hydrogen-bonding interactions. The reaction proceeds through concerted 9,10-addition with a Gibbs energy barrier of 25.5 kcal/mol, lower by more than 4 kcal/mol respect 1,4-additions in line with complete selectivity observed experimentally. The regioselectivity stands-on three key factors: (i) the higher electron density at the central ring, as indicated by the ESP analysis of substrate **1a**; (ii) the lower distortion energy loss along the transition state of the 9,10-addition, as revealed by the distortion/interaction analysis; and (iii) the greater HOMO-LUMO overlap observed through the secondary orbital interaction (SOI) analysis.

The analysis of the reaction catalysed by the bowl-shaped cage BS-$[Pd_6L_4]^{12+}$ reveals that the formation of the inclusion complex **1a**$\subset[Pd_6L_4]^{12+}$ is more favourable than **1b**$\subset BS$-$[Pd_6L_4]^{12+}$ driven by $\pi \cdots \pi$ stacking between the π-system of the cage wall and the anthracene aromatic rings. Both the 9,10- and 1,4-addition pathways show Gibbs energy barriers lower by more than 6 kcal/mol compared to bulk solution, proving enhanced catalytic efficiency; this was mostly attributed to NCIs that stabilize both intermediates and transition states. Moreover

($\mathbf{1a}$ + $\mathbf{1b}$)⊂BS-$[Pd_6L_4]^{12+}$ intermediates and TSs are further favoured by reducing the entropy and thermal contributions about 13 kcal/mol compared to the bulk solution. In coherence with the experiments, the 9,10-addition is still the most favoured pathway by 3.0 kcal/mol. Importantly, the encapsulated product **4** displays less binding energy compared to the ($\mathbf{1a}$ + $\mathbf{1b}$)⊂BS-$[Pd_6L_4]^{12+}$, thus facilitating the catalytic cycle by replacing the product with incoming reagents.

Concerning the octahedral cage, similar analyses drawn the conclusion that inclusion complex **1a**⊂O_h-$[Pd_6L_4]^{12+}$ is favoured over **1b**⊂O_h-$[Pd_6L_4]^{12+}$ as well as the formation of ($\mathbf{1a}$ + $\mathbf{1b}$)⊂O_h-$[Pd_6L_4]^{12+}$. However, out of the two 9,10- and 1,4-pathways, the confinement effect notably reduces only the Gibbs energy barrier of the 1,4-addition by 5.2 kcal/mol compared to bulk solution. This pathway is favoured by 1.2 kcal/mol, which reflects a predicted regioselectivity of 77%, closely aligned with the 100% observed experimentally. The switching in regioselectivity observed within the confined space of O_h-$[Pd_6L_4]^{12+}$ was rationalized accounting for: (i) the more contracted 1,4-addition TS (kinetic diameter of 13.13 vs 15.18 Å) indicating that both co-substrates are better encapsulated within cage compared to the 9,10-addition, in which **1a** is partially outside the cage; and (ii) the stronger $\pi \cdots \pi$ interaction and hydrogen bonding in 1,4-addition TS, as well as increased host-guest complementary as revealed by NCI and free volume analysis.

As previously commented in the binding section, Fujita, Takezawa and co-workers showed that the reaction rate for the hydrolysis of amides was significantly increased within octahedral $[Pd_6L_4]^{+12}$ metallocage *(50)*. In this section we will focus on the theoretical investigation performed by Pavan and co-workers on the reactivity itself (Fig. 10) *(51)*.

They compared three host-guest systems for the reactivity: (i) the amide guest alone inside the host (A⊂$[Pd_6L_4]^{12+}$), (ii) two amide guests inside the host (A_2⊂$[Pd_6L_4]^{12+}$), and (iii) the amide and a co-guest (phenanthrene) inside the host (A·P⊂$[Pd_6L_4]^{12+}$). The hydrolysis of the amide in these three systems was also compared with the reference process in solution (the amide alone in solution). The accelerations of ~26-fold, ~64-fold, and ~150-fold were computed for the A⊂$[Pd_6L_4]^{12+}$, A_2⊂$[Pd_6L_4]^{12+}$, and A·P⊂$[Pd_6L_4]^{12+}$ systems, respectively. They observed that the reactive conformer of the amide guest is more stabilized upon the encapsulation of the co-guest. They found that the enhanced reactivity observed experimentally within the cavity is controlled by the molecular crowding effects together with the host-guest complexation.

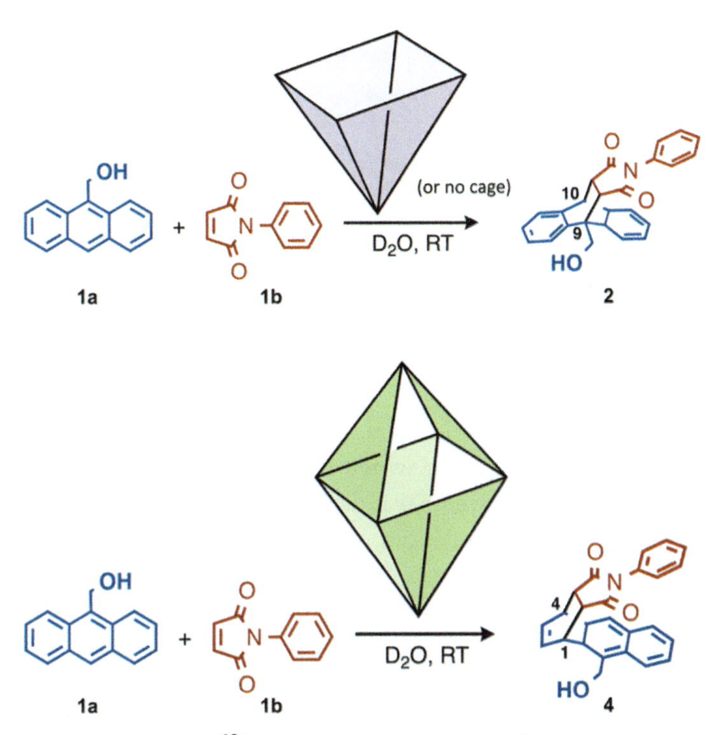

Fig. 10 Diel-Alder $[Pd_4L_4]^{12+}$ catalysed reactions and regioisomers products depending on the metallocage employed.

They also calculated the number of contacts between the carbonyl group of the amide guest and solvent molecules and observed that the trans isomer of the guest was slightly more accessible to the solvent relative to that of the cis isomer when the guests are in solution. Interestingly, this situation was switched for the A·P⊂$[Pd_6L_4]^{12+}$ system (the cis isomer of the amide was found to be more exposed to the solvent within the host compared to that of the trans isomer for the A·P⊂$[Pd_6L_4]^{12+}$).

The group of Klajn developed an octahedral $[Pd_6L_4]^{12+}$ metallocage that allow them controlling the isomerization process for encapsulated azobenzenes (Fig. 11) *(72)*. *Trans-cis* isomerization of azobenzenes inside such $[Pd_6L_4]^{12+}$metallocage, was investigated by Pavan and co-workers using classical MD and metadynamics *(73)*.

During the MD simulations, the substrates remained bound within the host. They observed that the host loses flexibility upon encapsulation. Moreover, the host must undergo structural rearrangements to incorporate the guest. The energy penalty for this guest-incorporation process was

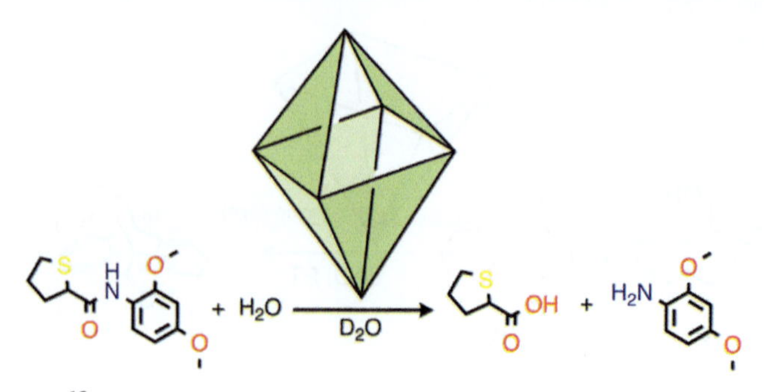

Fig. 11 $[Pd_4L_4]^{12+}$ catalysed amides hydrolysis reactions.

similar for both cis and trans isomers. The process inside the metallocage was simulated using a model in which the potential term for the central CN=NC dihedral of azobenzene was derived from QM calculations on the excited state. During the simulations, the trans-azobenzene guest (which is assumed to reach the excited state) undergoes spontaneous trans→cis isomerization. However, this is not enough to draw conclusions on whether the process takes place inside or outside the host. Therefore, they conducted additional metadynamics simulations on the guest encapsulation to obtain a complete picture of the transition mechanism and found that formation of the cis isomer is energetically favoured ($\Delta G_{binding}$ from −3.6 up to −7.9 kcal/mol) with slight lower activation barriers ($\Delta G_{encapsulation}^{\ddagger}$ about 0.3–2.8 kcal/mol). Overall, the Gibbs-energy cost associated with the host-guest complexation (molecular crowding inside the host) and the host-guest affinity were found to be the main factors governing these processes.

3.5 Pyramidal-shaped cages

A self-assembled tetrahedron metallocage, $K_{12}[Ga_4L_6]$, developed by the Raymond group has been applied as a nanoreactor for several chemical reactions including alkyl-alkyl reductive eliminations from high valent transition metal complexes (Au^{III} and Pt^{IV}), orthoformate hydrolysis, Nazarov cyclization, Prins cyclization, aza-Cope arrangement, … etc. *(74)*. (see Fig. 12).

The alkyl-alkyl reductive elimination from the Au(III) complex, $R_3PAuI(CH_3)_2$, was accelerated up to more than seven orders of magnitude when compared with the solution reaction (Fig. 13) *(74)*. The origin of this acceleration has been investigated computationally by two independent groups. The Teresa Head-Gordon group reported a DFT study on

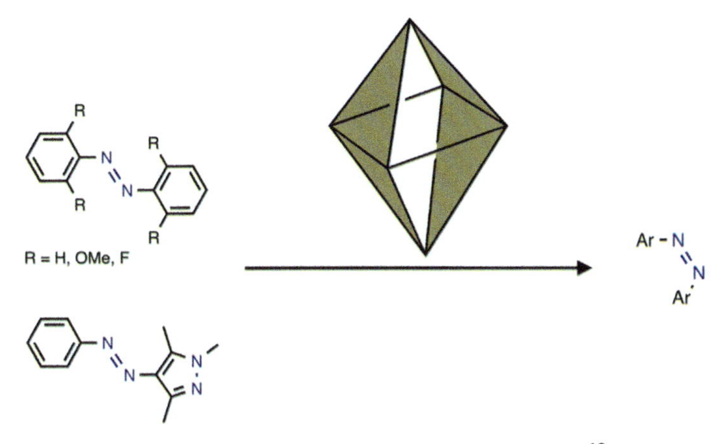

Fig. 12 cis-trans isomerization of azabenzenes inside the $[Pd_6L_4]^{12+}$ metallocage.

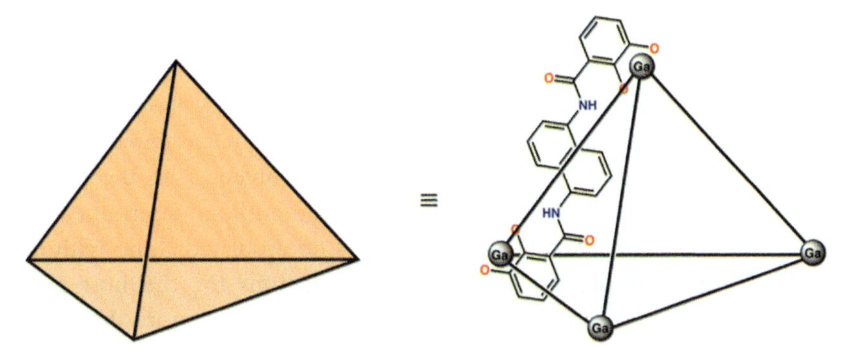

Fig. 13 Schematic representation of the $[Ga_4L_6]^{12-}$.

this process, ascertaining that the dissociated form of the Au(III) complex, $[(CH_3)_3PAu(CH_3)_2]^+$, is the resting state of the reaction *(75)*. Binding of this cation inside two self-assemblies, $[Ga_4L_6]^{12-}$ and $[Si_4L_6]^{8-}$ was analysed in vacuum and with a reaction field continuum solvent model. The analysis showed the preference for a more negatively charged host, as can be expected from electrostatic considerations. Moreover, the difference in electrostatics between these two hosts was also found to be the main factor for the transition state stabilization. When the continuum solvent model was not employed, the calculated energy barriers for the reductive elimination from $[(CH_3)_3PAu(CH_3)_2]^+$ inside $[Ga_4L_6]^{12-}$ and $[Si_4L_6]^{8-}$ with the explicit first solvation shell and counter ions were 5.5 and 24.4 kcal/mol, respectively. This difference of 18.9 kcal/mol in the barrier remained (~18.0 kcal/mol) after removing the explicit water solvent molecules inside the host, suggesting that the solvent water molecules inside these two hosts may have similar energetic contributions for this process. When the continuum solvent model was employed, the difference in the barrier decreased from 18.9 to ~8.0 kcal/mol.

The origin of the rate acceleration observed for the reaction in the $[Ga_4L_6]^{12-}$ was also analysed by the same group using ab initio metadynamics (AIMetaD) *(76)*. Two CVs, the distance between the carbon atoms forming C-C bond and the coordination number between the gold atom and two carbon atoms of the leaving methyl groups, were chosen for the AIMetaD calculations to obtain the Gibbs energy landscape. The calculated activation Gibbs energy barrier inside the $[Ga_4L_6]^{12-}$ was ~9 kcal/mol lower than that in solution. The electrostatic contribution to the barrier decrease was estimated to be ~5 kcal/mol as the electrostatic free energy of transition state stabilization, while the further ~50% of the transition state stabilization was attributed to the presence of the water molecule coordinated to the gold complex inside the $[Ga_4L_6]^{12-}$.

The origin of the rate acceleration on the same process was investigated by Ujaque and co-workers using DFT calculations in combination with classical MD simulations with explicit methanol solvent as in the experiment *(77)*. Using classical MD simulations, the authors determined the average number of solvent MeOH molecules encapsulated along with the Au(III) complex inside the $[Ga_4L_6]^{12-}$. Simulations for the host–guest complexes were performed in a cubic periodic box with more than 3500 explicit MeOH molecules and potassium counter ions to neutralize the system. The results obtained from the MD simulations guided the models selected for the DFT geometry optimizations. The DFT calculations for the complete host-guest system showed that in the presence of the metallocage, the Gibbs energy barrier for the reaction decreases by ~9 kcal/mol compared to that in solution. The overall decrease can be related to two different factors: microsolvation (removing solvent molecules around the Au(III) complex) and encapsulation (performing the reaction inside the cavity of the $[Ga_4L_6]^{12-}$).

Our research group also investigated the effects of the phosphine ligand of the Au(III) complex on the energy barrier by comparing $[(Me_3P)AuI(CH_3)_2]$ and $[Et_3PAuI(CH_3)_2]$ complexes inside the $[Ga_4L_6]^{12-}$ as well as in solution *(78)*. Experimentally, the reductive eliminations from these two Au(III) complexes have very similar reaction rates in solution but in the presence of the $[Ga_4L_6]^{12-}$, the difference in reaction rates between two processes is around two orders of magnitude. The MD simulations showed that for the Au(III) complex with the Me_3P ligand, there is in average one additional MeOH solvent molecule inside the $[Ga_4L_6]^{12-}$ compared to the Au(III) complex with the Et_3P ligand, indicating microsolvation inside the cavity of the $[Ga_4L_6]^{12-}$ is different for both Au(III) complexes. The DFT calculations showed that this difference is behind the experimental results since the microsolvation was found to be one of the factors responsible for the rate acceleration of the reaction. Overall, the results indicated that the encapsulated solvent molecules inside the $[Ga_4L_6]^{12-}$ can play a significant role in modifying the energy barrier of this process. The reductive elimination from a Pt(IV) complex, $[(Me_3P)_2PtI(CH_3)_3]$, encapsulated by the $[Ga_4L_6]^{12-}$ was investigated by the same authors *(79)*.

The reaction was modelled inside the $[Ga_4L_6]^{12-}$ and in solution in order to identify the origin of the observed rate acceleration. The MD simulations showed that there is no additional solvent MeOH molecule inside the $[Ga_4L_6]^{12-}$ when the complex, $[(Me_3P)_2Pt(MeOH)(CH_3)_3]^+$, is encapsulated. The DFT calculations showed that the Gibbs energy barrier for the reductive elimination decreases by ~7 kcal/mol in the $[Ga_4L_6]^{12-}$ compared to that in solution. Interestingly, the energy decomposition analysis showed that the microsolvation was not a significant factor, instead, the encapsulation was the main responsible factor for the rate acceleration observed experimentally. It is worth to note that even though the number of the encapsulated solvent molecule inside the $[Ga_4L_6]^{12-}$ is the same metal complexes ($[(Me_3P)_2Pt(MeOH)(CH_3)_3]^+$ and $[Et_3PAu(MeOH)(CH_3)_2]^+$) and the overall lowering of the energy barrier is comparable, the microsolvation and encapsulation effects can be significantly different depending on the nature of the metal complex (Au^{III} vs Pt^{IV}), resulting different origins for the observed rate accelerations. A possible explanation for the negligible microsolvation effect for the Pt(IV) complex compared to the Au(III) complex is that the Pt(IV) complex is an octahedron complex while the Au(III) complex is a square planar complex and therefore the Pt (IV) complex is more shielded from the surrounding solvent environments.

The effects of the metal substitution for the $[Ga_4L_6]^{12-}$ host on the reductive elimination from the Au(III) complex, $[(CH_3)_3PAu(CH_3)_2]^+$, were reported by Head-Gordon and co-workers group using AIMD calculations and electric field analyses in a subsequent work *(80)*, showing that replacing the four gallium atoms by four indium atoms lowered the activation energy barrier by ~3 kcal/mol for the same reaction in water solvent. The source of this lowering in the barrier was found to be a better arrangement of solvent water molecules around the metal centers of the $[In_4L_6]^{12-}$ compared to the $[Ga_4L_6]^{12-}$.

Another reaction catalysed by the $[Ga_4L_6]^{12-}$ is the hydrolysis of orthoformates in basic solution. The origin of this catalysis was studied by the Warshel group using the empirical valence bond and FEP approaches (Fig. 14) *(81)*. Their study revealed a significant stabilization of the hydronium ion, H_3O^+, inside the cavity of the $[Ga_4L_6]^{12-}$ host, resulting a condition with very low "local pH" for the reaction even in basic bulk solution and showed that the catalytic effect can be entirely attributed to the electrostatic contributions.

Enantioselectivity in the 3-aza-Cope rearrangement of allyl enammonium cations encapsulated in the $[Ga_4L_6]^{12-}$ was investigated by means of QM/MM calculations where the QM region includes the reactants and the metallocage, while the counterions were described at MM level *(82)*. The authors found a large difference in the energy barrier for the reaction pathways leading to the formation of R- and S-enantiomers. The former is calculated to be energetically lower than the latter along the reaction coordinate including the reactant states. The energy difference between the two reactants was ascribed to the host–guest shape complementarity. The origin of enantioselectivity is found in the different stability of the prochiral substrates in addition to the difference in the reaction barrier.

The Nazarov cyclization is also catalysed by the $[Ga_4L_6]^{12-}$; the Bergman and Raymond groups reported a mechanistic analysis of the Nazarov cyclization of 1,3-pentadienols encapsulated by the $[Ga_4L_6]^{12-}$ (Fig. 15) using a combination of experiments (kinetic and ^{18}O-incorporation studies) and calculations (DFT calculations for the reaction in solution) *(83)*.

Their analysis suggested that protonation of the alcohol substrate occurs along with its rapid and reversible binding, which is in agreement with the ease for the encapsulation of the hydronium ion, H_3O^+, inside the $[Ga_4L_6]^{12-}$ reported by the Warshel group in their study of the hydrolysis

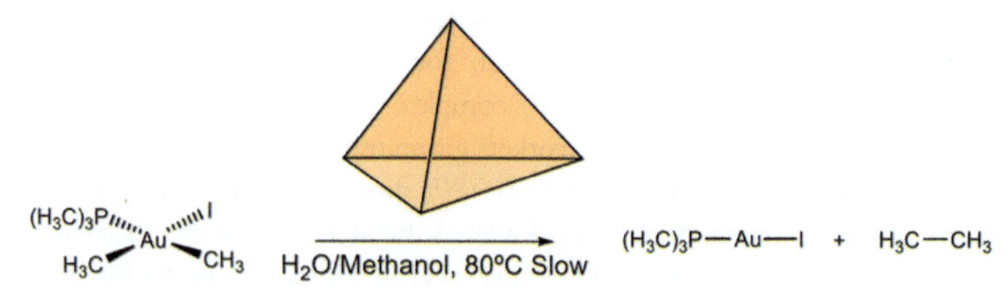

Fig. 14 $[Ga_4L_6]^{12-}$-catalysed alkyl-alkyl reductive elimination from $(CH_3)_3PAuI(CH_3)_2$ complex.

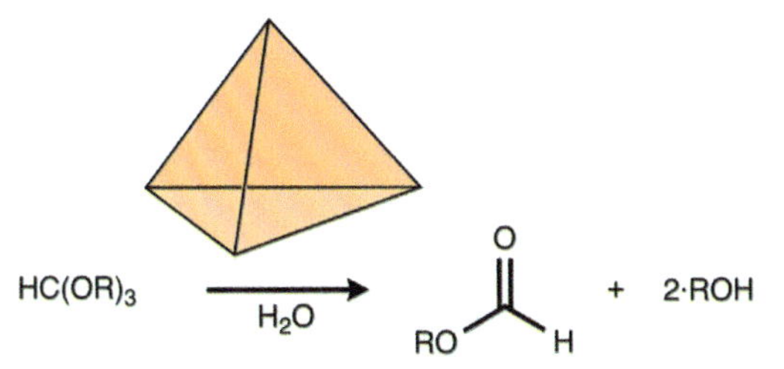

Fig. 15 $[Ga_4L_6]^{12-}$-catalysed hydrolysis of orthoformates.

of orthoformates. Overall, their results showed that for the uncatalyzed reaction, electrocyclization is the rate-determining step while for the catalysed reaction, both the water loss from protonated alcohol substrate and the electrocyclization are rate-limiting. To elaborate the origin of the catalysis and the effects of the host on each step of the reaction, two independent research papers investigated the process computationally in parallel using quantum chemical calculations and MD simulations; one identified and calculated the main interactions between the metallocage and the reactant *(84)*, whereas the other was able to obtain the Gibbs energy profile for the reaction in solution and inside the metallocage *(85)*. Both studies revealed that the main factor responsible for the catalysis was the change of the basicity of the alcohol substrate once encapsulated within the host.

As reported by Raymond, Bergman, Toste and co-workers *(86)*, the aza-Prins reaction between amine, **1c**, and formaldehyde, **2c**, follows different chemo-selectivity in bulk (formic acid) or in presence of the $[Ga_4L_6]^{12-}$ host in a solution of MeOH:H_2O with 1:1 ratio (Fig. 16).

1c and protonated **2c** react to produce alcohols or substituted piperidine products in bulk solution and within $[Ga_4L_6]^{12-}$, respectively. Xu and co-workers *(87)*, recently performed an integrated DFT/MM (B3LYP-D3(BJ):PM6D3) of the aza-Prins reaction in both the bulk solution and within the confined space of the $[Ga_4L_6]^{12-}$ anionic cage; an AIMD analysis (PBE, explicit H_2O solvent) was performed to select initial structure for DFT/MM analysis. The full-DFT analysis of the aza-Prins cyclization mechanism in bulk solution starts with the exergonic formation (ΔG = −38.2 kcal/mol) of the supramolecular adduct (**1c** + **2cH$^+$**). The following step is the dehydrative condensation through a H_2O-assisted six-membered ring TS with an associated Gibbs energy barrier of 23.7 kcal/mol, leading to an iminium ion intermediate. Low energy rearrangements to chairlike conformations can take place leading to two accessible intermediates with the relative positions of the C–C and C–N double bonds either in parallel or perpendicular disposition to each other. From these two intermediates, the new C-C bond formation and cyclization diverges leading, to two different carbocations which evolves toward alcohols **3c**, or substituted piperidine **4c**, after hydration or N-demetylation, respectively (Fig. 17).

The parallel-conformation pathway leading to the alcohol product resulted favoured by 3.8 kcal/mol in bulk solution in agreement with the chemo-selectivity observed experimentally. This energy difference is rationalised by the vertical-conformation pathways exhibiting greater steric repulsion due to shorter H–H distances, while the parallel-conformation arrangement enables stronger donor–acceptor orbital interactions. NCIs and distortion/interaction analysis over the selectivity-determining TSs pointed out that the selectivity observed into the confined space in mainly driven by: (i) stronger C–H···O hydrogen bonding and stabilizing C–H···π and π···π interactions, between the guest and host in the vertical-cyclization TS; (ii) larger distortion of the host, and steric H–H repulsions in the parallel-cyclization TS. (Fig. 18).

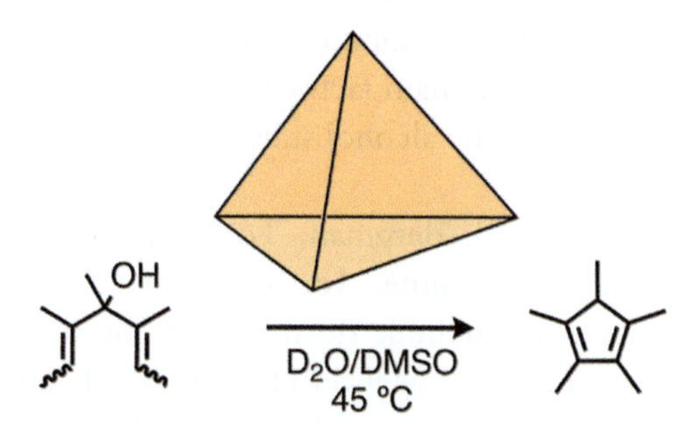

Fig. 16 $[Ga_4L_6]^{12-}$-catalysed Nazarov cyclization.

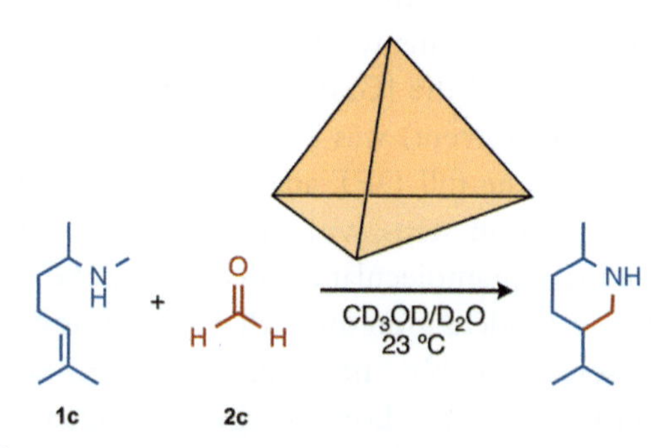

Fig. 17 $[Ga_4L_6]^{12-}$-catalysed aza-Prins reaction.

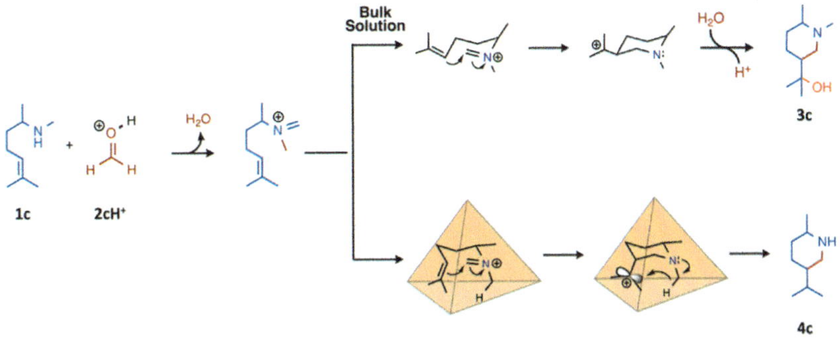

Fig. 18 Proposed mechanism for the $[Ga_4L_6]^{12-}$-catalysed aza-Prins reaction.

Prior the QM/MM mechanistic analysis, the authors ran exploratory AIMD simulations to analyse the Iminium⊂$[Ga_4L_6]^{12-}$ inclusion complex highlighting the positive region of the iminium co-substrate preferentially oriented toward the corner of the cage with C–C double stacked to the naphthalene ring of the cage's walls. Moreover, $3H_2O$ molecules were found inside the cage, binding to its corners in a 3-water chain. After geometry optimization of 24 snapshots extracted from the equilibrated trajectory the most table one was selected as the starting point for the mechanistic analysis. In contrast to the bulk solution, the QM/MM mechanistic analysis of the confined reaction, highlighted an energetic preference for the vertical-cyclization pathway by 4.2 kcal/mol assessing the predictive capabilities of the model.

Nitschke and co-workers reported that the anionic tetrahedral $[Fe_4L_6]^{4-}$ (with L = 4,4′-bis((pyridin-2-ylmethylene)amino)-[1,1′-biphenyl]−2,3′-disulfonate) enhances the reduction of small aromatic aldehydes to alcohols with high selectivity using the mild sodium cyanoborohydride reducing agent and neutral pH. Symes, Sproules and co-workers *(88)*, analysed by means of full-DFT (PB86-D3) calculations both furfural encapsulation and its reactivity propensity inside the MOC. Their studies first confirmed that furfural fits into the host forming the furfural⊂$[Fe_4L_6]^{4-}$ inclusion complex. Encapsulated furfural is significantly more stable than free furfural, and its orientation is influenced by NCIs within the cage. NCI analysis on the resulting geometry showed that the stabilisation of the guest is mainly driven by CH(furfural)···π(L) interactions, CO(furfural)···(L) hydrogen bonds and VdW contacts. Canonical Molecular Orbital analysis revealed that encapsulation stabilizes lowest unoccupied molecular orbital (LUMO) of furfural, suggesting enhanced electrophilic character of the carbonylic carbon.

4. Conclusions

Supramolecular chemistry stands as a robust discipline founded in stablishing molecular interactions beyond traditional covalent bonds. The arrival of synthetic hosts, notably MOCs, has broadened the horizons of supramolecular entities, offering diverse topologies and nanoscale architectures with a myriad of applications including molecular recognition, drug delivery and catalysis. Indeed, supramolecular catalysts exhibit remarkable efficiency and selectivity, sometimes comparable to those of enzymes.

In this chapter, we have surveyed computational investigations into rate accelerated processes facilitated by MOCs, shedding light on molecular binding processes and enhanced reactions within their cavities. While theoretical studies on these accelerated reactions are still rather limited, here it was shown that encapsulation process and reaction mechanism can be precisely studied using a proper multiscale approach. At this point one can envisage that the application of Machine Learning to supramolecular chemistry will profoundly affect this field, as many others. Nevertheless, unlike protein–ligand complexes, host–guest systems lack large training sets. Thus, the suitability of machine learning approaches needs to be accurately evaluated. At any rate, it is not within the focus of this chapter because there are no examples applied to MOCs yet.

For reliable insights into reactivity, where bonds are form and broken, QM-based methods are required. Because of the considerable size of these systems, conducting systematic studies at QM level of theory is quite challenging, though possible as shown in several examples, but QM/MM-based methods are frequently employed. The selection of structures to be computed is highly benefited by previous MD simulations. The examples described in the reactivity section shows that a proper selection of the QM or QM/MM methods allow an accurate analysis of the reaction mechanisms in both, bulk solution and within the cavities of the MOCs. The comparison among them is often quite instructive in understanding the origin of the rate acceleration.

Another crucial aspect is related to the guest release and encapsulation as well as the cage's flexibility during these events. Analysing these processes is also challenging because they involve dealing with a vast array of potential host-guest configurations; this is even aggravated when explicit solvent is considered. These processes have been mainly investigated employing MD simulations. Binding events takes place in a time scale going from milliseconds to seconds, a timeframe conventionally inaccessible to standard MD simulations; therefore, rare event sampling techniques like US, aMD, the APR method, and metadynamics have been employed. All of them have shown to be precise enough for describing these processes as well as for obtaining accurate binding energies.

Overall, although application of computational chemistry to supramolecular catalysis is still in an early stage, we firmly believe it holds a very promising future. This progress is likely to be driven by utilizing multiscaling methodologies. Employing QM, MM, and MD simulations, and integrating them effectively appears to be the most practical approach to tackle computational investigations into MOCs and their catalytic applications up to now. The use of MLPs will also advance the proficiency of the computational analysis.

Acknowledgements

Grants PID2020–116861GB-I00 and PID2023–150881NB-I00 funded by MCIN/AEI/ 10.13039/501100011033 are acknowledged.

References

1. a) Steed, J. W.; Atwood, J. L. *Supramolecular Chemistry;* Wiley, 2013; b) Lehn, J.-M. *Chem. Soc. Rev.* **2007,** *36*, 151–160; c) Atwood, J. L.; Gokel, G. W.; Barbour, L. J.; Rissanen, K.; Jayawickramarajah, J.; Wilson, A. J.; Dalgarno, S.; MacGillivray, L. R.; Glass, T. E.; Raston, C. *Comprehensive Supramolecular Chemistry II;* Elsevier, 2017.
2. a) Van Leeuwen, P. W. *Supramolecular Catalysis;* John Wiley & Sons, 2008; b) van Leeuwen, P. W.; Raynal, M. *Supramolecular Catalysis: New Directions and Developments;* John Wiley & Sons, 2022.
3. Hosseini, M. W.; Lehn, J.-M. *J. Chem. Soc., Chem. Commun.* **1991**, 451–453. https://doi.org/10.1039/C39910000451.
4. a) Rebilly, J.-N.; Colasson, B.; Bistri, O.; Over, D.; Reinaud, O. *Chem. Soc. Rev.* **2015,** *44*, 467–489; b) Meeuwissen, J.; Reek, J. N. H. *Nat. Chem* **2010,** *2*, 615–621.
5. a) Raynal, M.; Ballester, P.; Vidal-Ferran, A.; Van Leeuwen, P. W. N. M. *Chem. Soc. Rev.* **2014,** *43*, 1734–1787; b) Raynal, M.; Ballester, P.; Vidal-Ferran, A.; van Leeuwen, P. W. N. M. *Chem. Soc. Rev* **2014,** *43*, 1660–1733; c) Zhang, D.; Ronson, T. K.; Nitschke, J. R. *Acc. Chem. Res.* **2018,** *51*, 2423–2436; d) Brown, C. J.; Toste, F. D.; Bergman, R. G.; Raymond, K. N. *Chem. Rev.* **2015,** *115*, 3012–3035.
6. Grommet, A. B.; Feller, M.; Klajn, R. *Nat. Nanotechnol.* **2020,** *15*, 256–271.
7. Warshel, A. *J. Biol. Chem.* **1998,** *273*, 27035–27038.
8. a) Yoshizawa, M.; Klosterman, J. K.; Fujita, M. *Angew. Chem. Int. Ed.* **2009,** *48*, 3418–3438; b) Cook, T. R.; Stang, P. J. *Chem. Rev.* **2015,** *115*, 7001–7045; c) Saha, R.; Mondal, B.; Mukherjee, P. S. *Chem. Rev.* **2022,** *122*, 12244–12307; d) Holliday, B. J.; Mirkin, C. A. *Angew. Chem. Int. Ed.* **2001,** *40*, 2022–2043; e) Han, M.; Engelhard, D. M.; Clever, G. H. *Chem. Soc. Rev* **2014,** *43*, 1848–1860.
9. a) Cook, T. R.; Zheng, Y.-R.; Stang, P. J. *Chem. Rev.* **2013,** *113*, 734–777; b) Ballester, P.; Fujita, M.; Rebek, J. *Chem. Soc. Rev.* **2015,** *44*, 392–393; c) Ibáñez, S.; Poyatos, M.; Peris, E. *Acc. Chem. Res.* **2020,** *53*, 1401–1413; d) Ham, R.; Nielsen, C. J.; Pullen, S.; Reek, J. N. H. *Chem. Rev.* **2023,** *123*, 5225–5261; e) Pullen, S.; Clever, G. H. *Acc. Chem. Res* **2018,** *51*, 3052–3064.

10. a) Voloshin, Y.; Belaya, I.; Krämer, R. *The Encapsulation Phenomenon: Synthesis, Reactivity and Applications of Caged Ions and Molecules;* Springer, 2016; b) García-Simón, C.; Garcia-Borràs, M.; Gómez, L.; Parella, T.; Osuna, S.; Juanhuix, J.; Imaz, I.; Maspoch, D.; Costas, M.; Ribas, X. *Nat. Commun* **2014,** *5*, 5557; c) Li, J.-R.; Zhou, H.-C. *Nat. Chem* **2010,** *2*, 893–898; d) Morimoto, M.; Bierschenk, S. M.; Xia, K. T.; Bergman, R. G.; Raymond, K. N.; Toste, F. D. *Nat. Catal* **2020,** *3*, 969–984.
11. a) Shteinman, A. A. *Catalysts* **2023,** *13*, 415; b) Takezawa, H.; Fujita, M. *Bull. Chem. Soc. Jpn* **2021,** *94*, 2351–2369.
12. Eisenstein, O.; Ujaque, G.; Lledós, A. What Makes a Good (Computed) Energy Profile?. In *New Directions in the Modeling of Organometallic Reactions;* Lledós, A., Ujaque, G., Eds.; Springer International Publishing: Cham, 2020; pp. 1–38. https://doi.org/10.1007/3418_2020_57.
13. a) Chakraborty, D.; Chattaraj, P. K. *J. Comput. Chem* **2018,** *39*, 151–160; b) Pahima, E.; Zhang, Q.; Tiefenbacher, K.; Major, D. T. *J. Am. Chem. Soc* **2019,** *141*, 6234–6246; c) Daver, H.; Harvey, J. N.; Rebek, J.; Himo, F. *J. Am. Chem. Soc* **2017,** *139*, 15494–15503; d) Goehry, C.; Besora, M.; Maseras, F. *ACS Catal.* **2015,** *5*, 2445–2451; e) Kim, S. P.; Leach, A. G.; Houk, K. N. *J. Org. Chem* **2002,** *67*, 4250–4260; f) Li, W.-L.; Head-Gordon, T. *ACS Central Sci.* **2021,** 7, 72–80; g) López-Coll, R.; Álvarez-Yebra, R.; Feixas, F.; Lledó, A. *Chem. Eur. J* **2021,** *27*, 10099–10106; h) Daver, H.; Rebek, J., Jr.; Himo, F. *Chem. Eur. J.* **2020,** *26*, 10861–10870; i) Li, T.-R.; Huck, F.; Piccini, G.; Tiefenbacher, K. *Nat. Chem* **2022,** *14*, 985–994; j) Capelli, R.; Piccini, G. *J. Phys. Chem. C* **2024,** *128*, 635–641.
14. a) Sciortino, G.; Norjmaa, G.; Maréchal, J. D.; Ujaque, G. *Catalysis by Metal–Organic Cages: A Computational Perspective,. in Supramolecular Catalysis; Supramolecular Catalysis,* **2022,** pp. 271–285. https://doi.org/10.1002/9783527832033.ch19; b) Piskorz, T. K.; Martí-Centelles, V.; Young, T. A.; Lusby, P. J.; Duarte, F. *ACS Catal* **2022,** *12*, 5806–5826; c) Piskorz, T. K.; Martí-Centelles, V.; Spicer, R. L.; Duarte, F.; Lusby, P. J. *Chem. Sci.* **2023,** *14*, 11300–11331.
15. a) Turcani, L.; Tarzia, A.; Szczypiński, F. T.; Jelfs, K. E. *J. Phys. Chem* **2021,** *154*; b) Young, T. A.; Gheorghe, R.; Duarte, F. *J. Chem. Inf. Model.* **2020,** *60*, 3546–3557.
16. a) Atwood, J. *Inclusion Phenomena and Molecular Recognition;* Springer Science & Business Media, 2012; b) Mooibroek, T. J.; Scheiner, S.; Valkenier, H. *ChemPhysChem* **2021,** *22*, 433–434; c) Rebek, J., Ed.; *Angew. Chem. Int. Ed* **1990,** *29*, 245–255.
17. Houk, K. N.; Leach, A. G.; Kim, S. P.; Zhang, X. *Angew. Chem. Int. Ed.* **2003,** *42*, 4872–4897.
18. King, E.; Aitchison, E.; Li, H.; Luo, R. *Front. Mol. Biosci.* **2021,** *8*.
19. a) Anderson, A. M.; Manet, I.; Malanga, M.; Clemens, D. M.; Sadrerafi, K.; Piñeiro, Á.; García-Fandiño, R.; O'Connor, M. S. *Carbohydr. Polym.* **2024,** *323*, 121360; b) El-Barghouthi, M. I.; Bodoor, K.; Abuhasan, O. M.; Assaf, K. I.; Al Hourani, B. J.; Rawashdeh, A. M. M., *ACS Omega* **2022,** 7, 10729–10737; c) Arabzadeh, H.; Walker, B.; Sperling, J. M.; Acevedo, O.; Ren, P.; Yang, W.; Albrecht-Schönzart, T. E. *J. Phys. Chem. B* **2022,** *126*, 10721–10731.
20. a) Rizzi, A.; Murkli, S.; McNeill, J. N.; Yao, W.; Sullivan, M.; Gilson, M. K.; Chiu, M. W.; Isaacs, L.; Gibb, B. C.; Mobley, D. L.; Chodera, J. D. *J. Comput. -Aided Mol. Des* **2018,** *32*, 937–963; b) Shen, W.; Zhou, T.; Shi, X. *Nano Res.* **2023,** *16*, 13474–13497; c) Limongelli, V. *WIREs Comput. Mol. Sci*, 10, **2020**, e1455.
21. a) Parac, M.; Etinski, M.; Peric, M.; Grimme, S. *J. Chem. Theory Comput.* **2005,** *1*, 1110–1118; b) Young Lee, G.; Bay, K. L.; Houk, K. N. *Helv. Chim. Acta* **2019,** *102*, e1900032.
22. Chipot, C. *Annu. Rev. Biophys.* **2023,** *52*, 113–138.
23. McDonald, N. A.; Duffy, E. M.; Jorgensen, W. L. *J. Am. Chem. Soc.* **1998,** *120*, 5104–5111.

24. Pagadala, N. S.; Syed, K.; Tuszynski, J. *Biophys. Rev.* **2017,** *9*, 91–102.
25. Wang, Z.; Sun, H.; Yao, X.; Li, D.; Xu, L.; Li, Y.; Tian, S.; Hou, T. *Phys. Chem. Chem. Phys.* **2016,** *18*, 12964–12975.
26. Hansson, T.; Marelius, J.; Åqvist, J. *J. Comput. -Aided Mol. Des.* **1998,** *12*, 27–35.
27. a) Genheden, S.; Ryde, U. *J. Chem. Theory Comput.* **2011,** 7, 3768–3778; b) Genheden, S.; Ryde, U. *Expert Opin. Drug Discov.* **2015,** *10*, 449–461.
28. Homeyer, N.; Gohlke, H. *Mol. Inf.* **2012,** *31*, 114–122.
29. a) Laio, A.; Parrinello, M. *Proc. Natl. Acad. Sci. U. S. A* **2002,** *99*, 12562–12566; b) Bussi, G.; Laio, A. *Nat. Rev. Phys* **2020,** *2*, 200–212; c) Laio, A.; Gervasio, F. L. *Rep. Prog. Phys* **2008,** *71*, 126601.
30. Kästner, J. *WIREs Comput. Mol. Sci.* **2011,** *1*, 932–942.
31. Lee, T.-S.; Allen, B. K.; Giese, T. J.; Guo, Z.; Li, P.; Lin, C.; McGee, T. D., Jr.; Pearlman, D. A.; Radak, B. K.; Tao, Y.; Tsai, H.-C.; Xu, H.; Sherman, W.; York, D. M. *J. Chem. Inf. Model.* **2020,** *60*, 5595–5623.
32. a) Shen, C.; Hu, Y.; Wang, Z.; Zhang, X.; Zhong, H.; Wang, G.; Yao, X.; Xu, L.; Cao, D.; Hou, T. *Brief. Bioinform* **2020,** *22*, 497–514; b) Li, H.; Sze, K.-H.; Lu, G.; Ballester, P. J. *WIREs Comput. Mol. Sci* **2021,** *11*, e1478; c) Serillon, D.; Bo, C.; Barril, X. *J. Comput. -Aided Mol. Des* **2021,** *35*, 209–222; d) Xu, P.; Sattasathuchana, T.; Guidez, E.; Webb, S. P.; Montgomery, K.; Yasini, H.; Pedreira, I. F. M.; Gordon, M. S. *J. Phys. Chem.* **2021**, 154.
33. Kumar, S.; Rosenberg, J. M.; Bouzida, D.; Swendsen, R. H.; Kollman, P. A. *J. Comput. Chem.* **1992,** *13*, 1011–1021.
34. Hamelberg, D.; Mongan, J.; McCammon, J. A. *J. Phys. Chem.* **2004,** *120*, 11919–11929.
35. Juber, S.; Wingbermühle, S.; Nuernberger, P.; Clever, G. H.; Schäfer, L. V. *Phys. Chem. Chem. Phys.* **2021,** *23*, 7321–7332.
36. Han, M.; Michel, R.; He, B.; Chen, Y.-S.; Stalke, D.; John, M.; Clever, G. H. *Angew. Chem. Int. Ed.* **2013,** *52*, 1319–1323.
37. Li, R.-J.; Holstein, J. J.; Hiller, W. G.; Andréasson, J.; Clever, G. H. *J. Am. Chem. Soc.* **2019,** *141*, 2097–2103.
38. Li, R.-J.; Han, M.; Tessarolo, J.; Holstein, J. J.; Lübben, J.; Dittrich, B.; Volkmann, C.; Finze, M.; Jenne, C.; Clever, G. H. *ChemPhotoChem* **2019,** *3*, 378–383.
39. Artmann, K.; Li, R.-J.; Juber, S.; Benchimol, E.; Schäfer, L. V.; Clever, G. H.; Nuernberger, P. *Angew. Chem. Int. Ed.* **2022,** *61*, e202212112.
40. a) Platzek, A.; Juber, S.; Yurtseven, C.; Hasegawa, S.; Schneider, L.; Drechsler, C.; Ebbert, K. E.; Rudolf, R.; Yan, Q.-Q.; Holstein, J. J.; Schäfer, L. V.; Clever, G. H. *Angew. Chem. Int. Ed.* **2022,** *61*, e202209305; b) Juber, S.; Schäfer, L. V. **2023,** *25*, 29496–29505.
41. De Jong, D. H.; Schäfer, L. V.; De Vries, A. H.; Marrink, S. J.; Berendsen, H. J. C.; Grubmüller, H. *J. Comput. Chem.* **2011,** *32*, 1919–1928.
42. a) Prinz, J.-H.; Wu, H.; Sarich, M.; Keller, B.; Senne, M.; Held, M.; Chodera, J. D.; Schütte, C.; Noé, F. *J. Phys. Chem* **2011,** *134*; b) Husic, B. E.; Pande, V. S. *J. Am. Chem. Soc.* **2018,** *140*, 2386–2396.
43. Norjmaa, G.; Vidossich, P.; Maréchal, J.-D.; Ujaque, G. *J. Chem. Inf. Model.* **2021,** *61*, 4370–4381.
44. Yin, J.; Henriksen, N. M.; Slochower, D. R.; Gilson, M. K. *J. Comput. -Aided Mol. Des.* **2017,** *31*, 133–145.
45. Torrie, G. M.; Valleau, J. P. *J. Comput. Phys.* **1977,** *23*, 187–199.
46. Sgarlata, C.; Mugridge, J. S.; Pluth, M. D.; Zito, V.; Arena, G.; Raymond, K. N. *Chem. Eur. J.* **2017,** *23*, 16813–16818.
47. García-Simón, C.; Colomban, C.; Çetin, Y. A.; Gimeno, A.; Pujals, M.; Ubasart, E.; Fuertes-Espinosa, C.; Asad, K.; Chronakis, N.; Costas, M.; Jiménez-Barbero, J.; Feixas, F.; Ribas, X. *J. Am. Chem. Soc.* **2020,** *142*, 16051–16063.

48. Frederix, P. W. J. M.; Patmanidis, I.; Marrink, S. J. *Chem. Soc. Rev.* **2018,** *47*, 3470–3489.
49. Miao, Y.; Feixas, F.; Eun, C.; McCammon, J. A. *J. Comput. Chem.* **2015,** *36*, 1536–1549.
50. Takezawa, H.; Shitozawa, K.; Fujita, M. *Nat. Chem.* **2020,** *12*, 574–578.
51. Delle Piane, M.; Pesce, L.; Cioni, M.; Pavan, G. M. *Chem. Sci.* **2022,** *13*, 11232–11245.
52. Koch, W.; Holthausen, M. C. *A Chemist's Guide to Density Functional Theory;* John Wiley & Sons, 2015.
53. Lledós, A.; Ujaque, G. *Topics in Organometallic Chemistry: New Direc-tions in the Modeling of Organometallic Reactions;* Springer Nature: Switzerland, AG, 2020.
54. Tarzia, A.; Jelfs, K. E. *Chem. Commun.* **2022,** *58*, 3717–3730.
55. Harvey, J. N.; Himo, F.; Maseras, F.; Perrin, L. *ACS Catal* **2019,** *9*, 6803–6813.
56. Tantillo, D. J. *Acc. Chem. Res* **2016,** *49*, 1079.
57. Chung, L. W.; Sameera, W. M. C.; Ramozzi, R.; Page, A. J.; Hatanaka, M.; Petrova, G. P.; Harris, T. V.; Li, X.; Ke, Z.; Liu, F.; Li, H.-B.; Ding, L.; Morokuma, K. *Chem. Rev.* **2015,** *115*, 5678–5796.
58. Grimme, S.; Bannwarth, C.; Shushkov, P. *J. Chem. Theory Comput.* **2017,** *13*, 1989–2009.
59. a) Thiel, W. *WIREs Comput. Mol. Sci.* **2014,** *4*, 145–157; b) Christensen, A. S.; Kubař, T.; Cui, Q.; Elstner, M. *Chem. Rev* **2016,** *116*, 5301–5337.
60. a) Li, P.; Merz, K. M. *Chem. Rev.* **2017,** *117*, 1564–1686; b) Seminario, J. M. *Int. J. Quantum Chem* **1996,** *60*, 1271–1277; c) Li, P.; Merz, K. M. *J. Chem. Inf. Model* **2016,** *56*, 599–604; d) Zheng, S.; Tang, Q.; He, J.; Du, S.; Xu, S.; Wang, C.; Xu, Y.; Lin, F. *J. Chem. Inf. Model* **2016,** *56*, 811–818.
61. Yang, Y.; Zhang, S.; Ranasinghe, K.; Isayev, O.; Roitberg, A. *ChemRxiv* **2023.** https://doi.org/10.26434/chemrxiv-2023-x82fz.
62. Friederich, P.; dos Passos Gomes, G.; De Bin, R.; Aspuru-Guzik, A.; Balcells, D. *Chem. Sci.* **2020,** *11*, 4584–4601.
63. Tomasi, J.; Mennucci, B.; Cammi, R. *Chem. Rev.* **2005,** *105*, 2999–3094.
64. Marenich, A. V.; Cramer, C. J.; Truhlar, D. G. *J. Phys. Chem. B* **2009,** *113*, 6378–6396.
65. Klamt, A.; Schüürmann, G. *J. Chem. Soc., Perkin Trans.* **1993**, 799–805. https://doi.org/10.1039/P29930000799.
66. Young, T. A.; Martí-Centelles, V.; Wang, J.; Lusby, P. J.; Duarte, F. *J. Am. Chem. Soc.* **2020,** *142*, 1300–1310.
67. Duarte, F.; Bauer, P.; Barrozo, A.; Amrein, B. A.; Purg, M.; Åqvist, J.; Kamerlin, S. C. L. *J. Phys. Chem. B* **2014,** *118*, 4351–4362.
68. Wang, J.; Young, T. A.; Duarte, F.; Lusby, P. J. *J. Am. Chem. Soc.* **2020,** *142*, 17743–17750.
69. Spicer, R. L.; Stergiou, A. D.; Young, T. A.; Duarte, F.; Symes, M. D.; Lusby, P. J. *J. Am. Chem. Soc.* **2020,** *142*, 2134–2139.
70. Yoshizawa, M.; Tamura, M.; Fujita, M. *Science* **2006,** *312*, 251–254.
71. Mao, X.-R.; Wang, Q.; Zhuo, S.-P.; Xu, L.-P. *Inorg. Chem.* **2023,** *62*, 4330–4340.
72. Samanta, D.; Gemen, J.; Chu, Z.; Diskin-Posner, Y.; Shimon, L. J. W.; Klajn, R. *Proc. Natl. Acad. Sci. U. S. A* **2018,** *115*, 9379–9384.
73. Pesce, L.; Perego, C.; Grommet, A. B.; Klajn, R.; Pavan, G. M. *J. Am. Chem. Soc.* **2020,** *142*, 9792–9802.
74. Kaphan, D. M.; Levin, M. D.; Bergman, R. G.; Raymond, K. N.; Toste, F. D. *Science* **2015,** *350*, 1235–1238.
75. Vaissier Welborn, V.; Head-Gordon, T. *J. Phys. Chem. Lett.* **2018,** *9*, 3814–3818.
76. Welborn, V. V.; Li, W.-L.; Head-Gordon, T. *Nat. Commun.* **2020,** *11*, 415.

77. Norjmaa, G.; Maréchal, J.-D.; Ujaque, G. *J. Am. Chem. Soc.* **2019,** *141*, 13114–13123.
78. Norjmaa, G.; Maréchal, J.-D.; Ujaque, G. *Chem. Eur. J.* **2020,** *26*, 6988–6992.
79. Norjmaa, G.; Maréchal, J.-D.; Ujaque, G. *Chem. Eur. J.* **2021,** *27*, 15973–15980.
80. Li, W.-L.; Hao, H.; Head-Gordon, T. *ACS Catal.* **2022,** *12*, 3782–3788.
81. Frushicheva, M. P.; Mukherjee, S.; Warshel, A. *J. Phys. Chem. B* **2012,** *116*, 13353–13360.
82. Ootani, Y.; Akinaga, Y.; Nakajima, T. *J. Comput. Chem.* **2015,** *36*, 459–466.
83. Hastings, C. J.; Bergman, R. G.; Raymond, K. N. *Chem. Eur. J.* **2014,** *20*, 3966–3973.
84. Nguyen, Q. N. N.; Xia, K. T.; Zhang, Y.; Chen, N.; Morimoto, M.; Pei, X.; Ha, Y.; Guo, J.; Yang, W.; Wang, L.-P.; Bergman, R. G.; Raymond, K. N.; Toste, F. D.; Tantillo, D. J. *J. Am. Chem. Soc.* **2022,** *144*, 11413–11424.
85. Norjmaa, G.; Himo, F.; Maréchal, J.-D.; Ujaque, G. *Chem. Eur. J.* **2022,** *28*, e202201792.
86. Kaphan, D. M.; Toste, F. D.; Bergman, R. G.; Raymond, K. N. *J. Am. Chem. Soc.* **2015,** *137*, 9202–9205.
87. Li, N.; Wang, Q.; Zhuo, S.; Xu, L.-P. *ACS Catal.* **2023,** *13*, 10531–10540.
88. Paul, A.; Shipman, M. A.; Onabule, D. Y.; Sproules, S.; Symes, M. D. *Chem. Sci.* **2021,** *12*, 5082–5090.

About the authors

Mercè Alemany-Chavarria is a Ph.D. student in Chemistry since 2022 from Universitat Autònoma de Barcelona (UAB). With a bachelor in Biochemistry and amaster in Bioinformatics, she has worked in IT consulting as DevOps and as abioinformatician in Vall d'Hebron Oncology Institute (VHIO) in Barcelona. Currently she is doing her Ph.D. studies in Chemistry, using computational tools to research binding processes in both organometallic supramolecular systems and metalloproteins.

Gantulga Norjmaa obtained his Ph.D. in Chemistry from the Universitat Autònoma de Barcelona (UAB) in 2021 under the direction of Profs. G. Ujaque and J.-D. Maréchal and then he joined the research group of Prof. F. Himo at the Stockholm University as a "Wenner-Gren" post-doctoral fellow. He is currently working as a postdoctoral researcher with Prof. G. Ujaque at the UAB. His main research interests are the use of computational methods for understanding chemical and biological processes. He received several awards including the Extraordinary Ph.D. award from the UAB and the Best Publication award (predoctoral category) from the Theoretical and Computational Chemistry Division of the Spanish Royal Society of Chemistry.

Giuseppe Sciortino obtained in 2019 his double Ph.D. in chemistry from both the Autonomous University of Barcelona (UAB, Spain) and the University of Sassari (UniSS,Italy), under the supervision of Professors Jean-Didier Maréchal and Eugenio Garribba. After one year of UniSS-UAB post-doctoral research, he was awarded a "Juan de la Cierva" Fellowship at the Institute of Chemical Research of Catalonia (Spain), working in the group of Professor Feliu Maseras. In 2023 he returned to UAB with a permanent position of associate professor. His research interests include the computational design and modeling of organometallic homogenous catalysts and artificial metalloenzymes, with a particular focus on asymmetric processes and complex multimetallic and photocatalytic systems.

Gregori Ujaque obtained his Ph.D. in Chemistry from the Universitat Autònoma de Barcelona (UAB) in 1999 under supervision of Profs. A. Lledós and F. Maseras. He did apostdoctoral stay at UCLA with Prof. K.N. Houk and then he returned to UAB bymeans of the "Ramon y Cajal" program, obtaining a permanent position in 2007. His main interests are the application of computational methods to understanding chemical reactivity and catalysis. He leads the work on catalytic processes as cross-coupling, hydroamination or Au-catalyzed reactions, among others. His interests have extended to modeling supramolecular catalysis in confined spaces. From 2018-2020 he served as associated editor of the journal "Anales de Química". Since 2020 he is the President of the Catalan Chemical Society (SCQ).

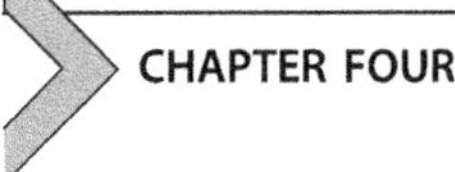

CHAPTER FOUR

Computational modeling of the epoxidation of alkenes with hydrogen peroxide catalyzed by transition metal-substituted polyoxometalates

Albert Solé-Daura[a,*] and Jorge J. Carbó[b,*]

[a]Institute of Chemical Research of Catalonia (ICIQ-CERCA), The Barcelona Institute of Science and Technology, Tarragona, Spain

[b]Departament de Química Física i Inorgaǹica, Universitat Rovira i Virgili, Tarragona, Spain

*Corresponding authors. e-mail address: asole@iciq.es; j.carbo@urv.cat

Contents

Abstract

The catalytic epoxidation of alkenes using hydrogen peroxide (H_2O_2) as a terminal oxidant is commonly referred to as a 'green oxidation' reaction, providing access to synthetically relevant epoxides while generating water s the sole byproduct. Polyoxometalates (POMs) are discrete polynuclear metal-oxo clusters, which can accommodate other transition metals in their structure. This results in transition metal-substituted POMs capable of activating H_2O_2 heterolytically for subsequent alkene epoxidation in a highly active and selective catalytic manner. This book chapter delves into nearly twenty years of computational research in the field of POM-catalyzed alkene epoxidation with H_2O_2. In particular, it focuses on how computational chemistry has contributed to the characterization of

Advances in Catalysis, Volume 75

ISSN 0360-0564, https://doi.org/10.1016/bs.acat.2024.08.003

reaction mechanisms at the atomic level and to the understanding of the factors affecting catalytic parameters such as activity and selectivity, which are crucial for the development of new catalysts with enhanced performance.

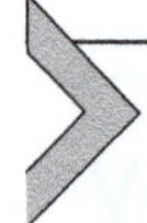

1. Introduction

1.1 Epoxidation of alkenes with hydrogen peroxide

The catalytic epoxidation of alkenes using hydrogen peroxide (H_2O_2) is a process of both academic and industrial interest. This process provides access to synthetically valuable compounds, i.e. epoxides, using H_2O_2 as a green and environmentally-friendly terminal oxidant, which ideally, gives water as the sole byproduct (Scheme 1). Epoxides can be regarded as synthetically relevant intermediates, key to the preparation of fine chemicals, epoxy resins, polymers, cosmetics or lubricants. On these grounds, there is a long-standing and ever-growing interest in producing them through green and sustainable pathways.

At the industrial scale, the heterogeneous titanium silicalite-1 (TS-1) catalyst has been largely used because it is a highly effective and stable for selective oxidations of small organic substrates, including the epoxidation of linear olefins with H_2O_2 as terminal oxidant *(1)*. To achieve the selective oxidation of bulky organic molecules with environmentally friendly oxidants, research activity directed to the synthesis of mesoporous metalsilicates, comprising Ti-containing materials as well as other transition metals such as niobium and zirconium *(2–5)*. Crystalline Metal-Organic Frameworks have been also applied to this purpose, as they constitute a family of highly tunable heterogeneous platforms that allow precise control over pore dimensions *(6,7)*.

Owing to the structural complexity of Ti-silicalite catalysts, the nature of its active sites has been a matter of intense debate in the literature and therefore, obtaining details about the underlying epoxidation mechanism at the atomic level has been a challenging task for decades. Specifically, both mononuclear and dinuclear sites, coordinating either peroxo or hydroperoxo ligands have been proposed to play the role of catalytically active species *(8–14)*. Recently, using spectroscopic and microscopic techniques,

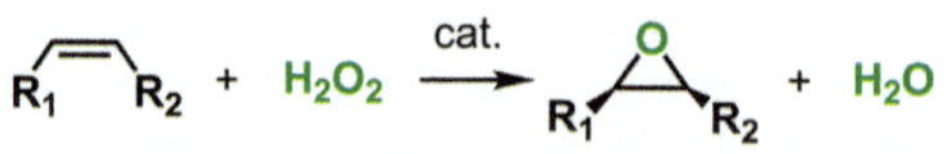

$$R_1\text{CH=CH}R_2 + H_2O_2 \xrightarrow{\text{cat.}} R_1\text{-epoxide-}R_2 + H_2O$$

Scheme 1 Equation for the catalytic epoxidation of alkenes with H_2O_2.

Copéret and coworkers provided evidence for the formation of bridging hydroperoxo groups connecting two neighboring Ti centers upon exposing TS-1 to H_2O_2 *(14)*. Moreover, Copéret and coworkers also characterized the mechanism responsible for heterolytic H_2O_2 activation and subsequent alkene epoxidation by the proposed catalytic sites of TS-1 by means of DFT calculations, which is summarized in Scheme 2. Following a heterolytic activation of H_2O_2 on Ti(IV) sites to form an active bridged hydroperoxo species, the electrophilic transfer of the more accessible O atom of the hydroperoxo group to the double bond of the alkene gives access to the epoxide product. This process involves the attack of the to the $\pi_{C=C}$ molecular orbital of the alkene to the empty σ^*_{O-O} of the oxidant, which takes place through a concerted spiro-like transition state, whereby the formation of both C–O bonds occurs concomitantly with the cleavage of the O–O bond. In TS-1, this was found to be assisted by a neighboring bridging O of the matrix, which abstracts the proton from the terminal oxygen of the hydroperoxo ligand, finally giving it back to the bridging Ti-O-Ti site after the O-transfer step and before regenerating the active form of the catalyst through the incorporation of a new molecule of H_2O_2. Despite subtle differences, the general nature of this step-wise mechanism is in line with previous proposals inferred from studies on model systems of TS-1 *(8,10,12,13)*, as well as on other homogeneous Ti-based catalysts *(15)*.

As discussed in following sections, this overall mechanistic picture also applies to polyoxometalate-based catalysts. However, more mechanistic subtleties regarding the real nature (hydroperoxo or peroxo) of the epoxidizing species or the position of the O center to be transferred to the

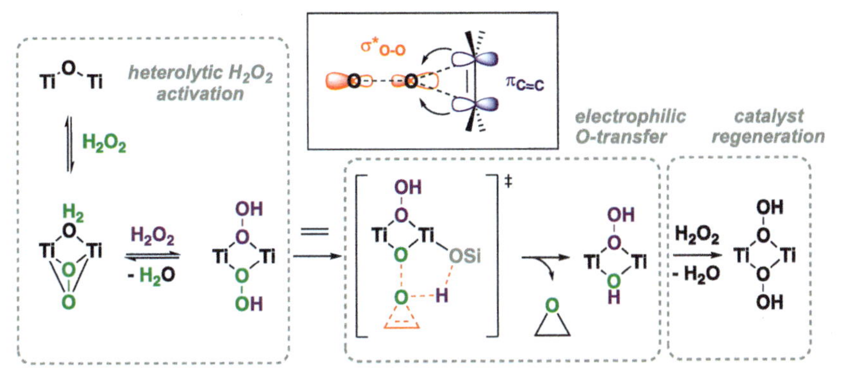

Scheme 2 Proposed reaction mechanism for the epoxidation of alkenes with H_2O_2 catalyzed by TS-1 *(14)*.

alkene have been reported depending on the nature of the catalytically active metal center and the molecular structure of the catalyst.

Beyond the mechanistic intricacies presented above, the catalytic activation of H_2O_2 for subsequent alkene epoxidation also faces selectivity issues, related to the unproductive decomposition H_2O_2, being catalyzed by the same catalysts that promote the epoxidation of alkenes *(16)*. This competing process is acknowledged to form radical intermediates such as $OH^{\bullet}$, $O_2^{\bullet-}$, $HO_2^{\bullet}$, or singlet oxygen *(17–24)* which can interact with olefinic substrates yielding the so-called "homolytic products" (Scheme 3), thus compromising not only the selectivity of the reaction, but also the efficiency of oxidant utilization *(1,25)*.

Hence, improving the efficiency and atomic economy of this process requires mechanistic knowledge not only about alkene epoxidation pathways but also about the competing side reactions. Understanding how these processes occur at the atomic level allows for a precise assessment of the factors that influence the activity and the selectivity, which jointly enable the development of clear catalyst design rules to improve catalytic performances. In this regard, polyoxometalates (POMs) have been extensively used as tractable molecular models for single-site and metal-oxide heterogeneous catalysts to allow fundamental understanding of the mechanism of epoxidation reactions and the individual factors affecting their activity and selectivity. However, these turned out being highly efficient and selective homogeneous catalysts *per se*.

1.2 Polyoxometalates as selective catalysts for alkene epoxidation with H_2O_2

POMs are polynuclear metal-oxide clusters built up from early transition-metal atoms such as W, Mo or V in their highest oxidation state *(26,27)*. Within the structure of POMs, metal ions are typically found in a distorted octahedral environment and link to one another through bridging oxo

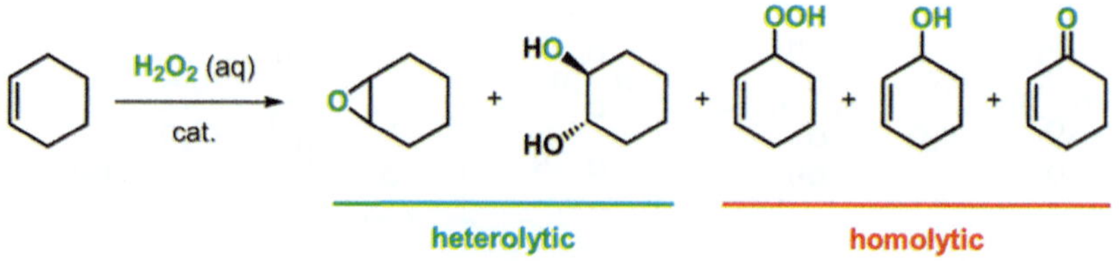

Scheme 3 Oxidation products stemming from the interaction between alkenes and metal-activated H_2O_2, following either heterolytic or homolytic pathways.

groups. Depending on the experimental conditions used for the synthesis of POMs (pH, concentration of precursors, ionic strength, presence/absence of main-group elements salts, etc.), metal-oxo units can self-assemble in different fashions, leading to well-established nanometric POM structures differing in nuclearity, size, shape or global negative charge. Fig. 1 depicts the structures of the Lindqvist and Keggin POMs, which constitute the basic scaffolds of the most representative POM-based catalysts used in alkene epoxidation.

The POM structures depicted in Fig. 1 represent "plenary structures", in which the metal-oxo groups self-assemble to form a closed structure with no room for coordinating any other metal ion. However, synthetic strategies have been developed to isolate metastable "lacunary" species, in which one or more metal units from plenary POMs are missing, as schematically illustrated in Fig. 2 *(28)*. Lacunary polyoxotungstates have proven active in the catalytic epoxidation of alkenes with H_2O_2, triggering early computational studies on this reaction, as discussed in Section 2. Nevertheless, lacunary POMs are highly negatively-charged species, which can be regarded as potential chelating ligands for catalytically active transition

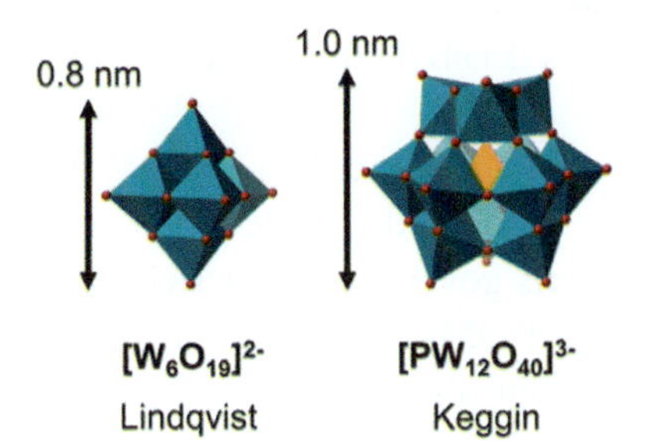

Fig. 1 Polyhedral representation of some archetypal POM structures.

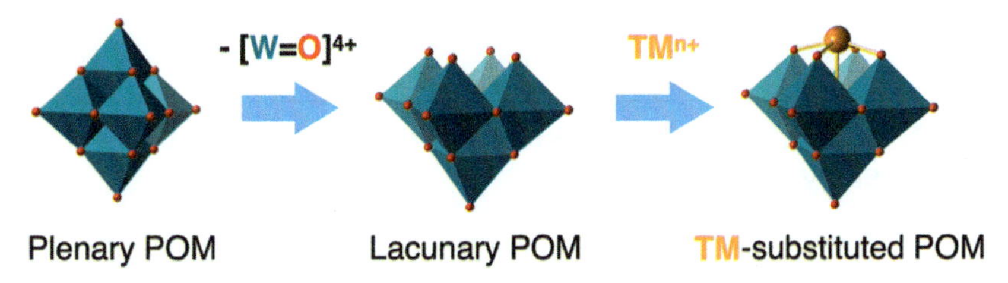

Fig. 2 Conceptual representation of the replacement of an addenda ion from a plenary POM structure by another TM ion.

metal (TM) ions. Hence, the combination of lacunary POMs with TM-based salts leads to the formation of TM-substituted POMs (Fig. 2), conferring them a broad applicability in catalysis. In particular, metal-substitution strategies began to be investigated in the field of oxidation catalysis and paved the way for a vast and continually-increasing number of experimental and computational studies on this topic *(1,25,29)*.

Such broad tunability enables the design and preparation of POM-based catalysts with tailored properties by varying the nature of the embedded metal ion, as exemplified in Sections 3 and 4. So far, V(V)-, Ti(IV)-, Nb(V)- and Zr (IV) substituted POMs have been successfully applied to the catalytic oxidation of alkenes with H_2O_2 and simultaneously explored from a computational standpoint. Following the trail of early studies detailed in Section 2, the vast majority of the computational studies on POM-catalyzed alkene epoxidation were carried out at the B3LYP/double-ζ+polarization level of theory. Upon the incorporation of solvent effects through implicit solvent models, this computational protocol has been consistently adopted over the years as a standard strategy to model epoxidation reactions with POM-based catalysts. Notably, the application of computational methods is key to obtain mechanistic insights that are otherwise challenging to obtain at the experimental level, providing a full characterization of elusive metastable intermediates and transition states (TSs) as well as details on the molecular features that influence their stability.

The book chapter aims to provide a critical overview of the fundamental understanding provided by computational studies on the mechanism and the factors governing the activity and selectivity of alkene epoxidation catalyzed by POMs. These complexes are not only interesting homogeneous catalysts, but also, they are tractable molecular models of solid catalysts such as MOF and single-site catalysts that enable detailed experimental characterization. Here, we focus on the computational studies backed by experimental results, ensuring the consistency of the atomistic learnings derived from calculations.

2. Epoxidation of alkenes by tungstate structures: early studies

Early computational studies by Musaev and coworkers on the epoxidation of alkenes with H_2O_2 catalyzed by POMs focused on the lacunary Keggin-type silicodecatungstate $[\gamma\text{-}(SiO_4)W_{10}O_{32}H_4]^{4-}$ (represented in Fig. 3A) *(30,31)*,

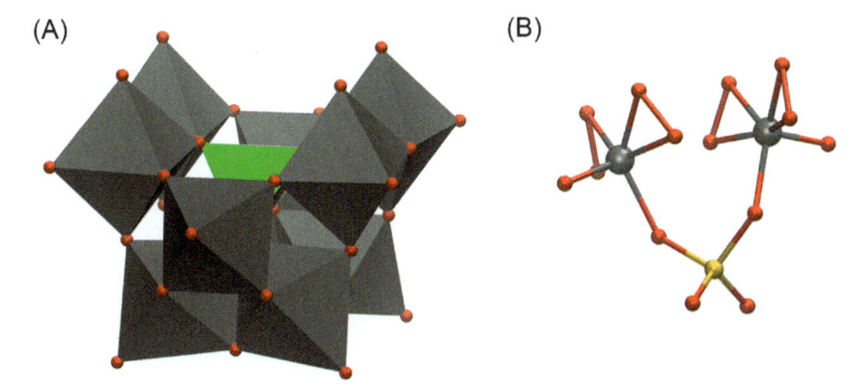

Fig. 3 Molecular structures of tungstate anions studied computationally for alkene epoxidation: (A) $[\gamma\text{-}(SiO_4)W_{10}O_{32}H_4]^{4-}$ silicodecatungstate, in which H atoms have been omitted given that their location has been a matter of debate (see main text); and (B) dinuclear peroxotungstate $[SeO_4\{WO(O_2)_2\}_2]^{2-}$.

which had been experimentally reported to catalyze the epoxidation of alkenes with high selectivity, stereospecificity and nearly total H_2O_2 utilization efficiency *(32,33)*. This lacunary POM possess structural features, which explain its activity compared with other tungstates such as $TBA_6[As_2W_{21}O_{67}(H_2O)]$ that are not active in alkene epoxidation *(34)*. The tungstate atoms at the lacuna contain hydroxo and aqua ligands that are more reactive towards the incoming H_2O_2 reactant. DFT calculations at the B3LYP/double-ζ level were firstly performed on a model system of the tungsten sites at the lacuna, consisting of an hexacoordinated W(VI) ion bearing an oxo group, four hydroxyl groups and an aqua ligand; then transitioning to a complete model of the POM. The IEF-PCM implicit solvent model was used to account for the dielectric effects of heptane, benzene, and acetonitrile solvents.

This computational investigation proposed the mechanistic picture given in Scheme 4, consisting of two distinct steps: (i) the heterolytic activation of H_2O_2 on a metal-hydroxo (TM-OH) group yielding a metal-hydroperoxo species (TM-OOH) in addition to a water molecule; and (ii) the cleavage of the O–O bond triggered by the nucleophilic attack of the olefinic substrate to give the epoxide product while regenerating the catalyst in its TM-OH form *(30,35)* While the first step was predicted to proceed smoothly, with electronic energy barriers of less than 5 kcal/mol in benzene, the electrophilic oxygen transfer to the alkene was proposed to be the rate-determining step of the whole process. For the latter, DFT calculations estimated a barrier of almost 40 kcal/mol, even though the height

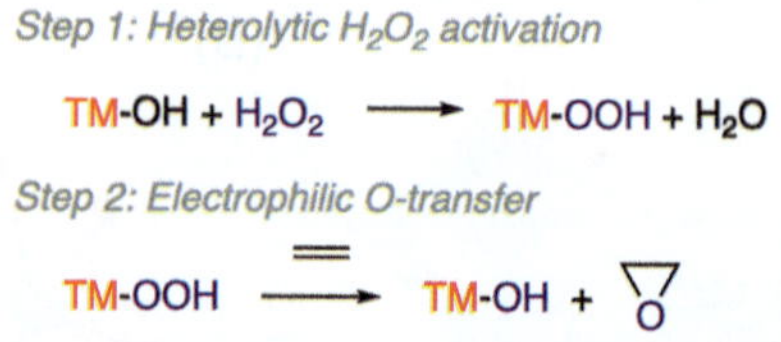

Step 1: Heterolytic H_2O_2 activation

$$\text{TM-OH} + H_2O_2 \longrightarrow \text{TM-OOH} + H_2O$$

Step 2: Electrophilic O-transfer

$$\text{TM-OOH} \xrightarrow{=} \text{TM-OH} + \text{epoxide}$$

Scheme 4 General mechanism of alknene epoxidation with H_2O_2 catalyzed by POMs.

of this barrier was found to decrease by *ca.* 12 kcal/mol upon the incorporation of explicit $N(CH_3)_4^+$ counter cations into the computational models. Conspicuously, this work evidences the central role of protonated oxygen sites at the POM surface. Particularly, these are essential to enable the heterolytic activation of the oxidant by being involved in the formation of an aqua ligand, which can be easily displaced from the coordination sphere of the metal to allow the coordination of the hydroperoxo group.

The formation of a TM-OO peroxo intermediate, and the possible O-transfer to the alkene from the latter were ruled out on the basis that such a highly reactive species would abstract protons from neighboring hydroxo groups *(30)*. Nevertheless, DFT calculations on Se-containing di- and tetra-nuclear peroxotungstates proposed an O-transfer mechanism involving a peroxo group *(36–38)*. Specifically, barriers of c. 20–24 kcal/mol were reported for nonfunctionalized alkenes, decreasing to 13–14 kcal/mol for allylic alcohols due to the stabilization of the TS structure provided by an intermolecular hydrogen bond between the alcohol function of the substrate and basic O sites of the catalyst. As we will discuss along the chapter the preference for the hydroperoxo or peroxo path depends on the metal nature, and they could be competitive for a given TM.

Shortly after the work by Musaev and coworkers *(30)* came out, the structural and catalytic features of $[\gamma\text{-}(SiO_4)W_{10}O_{32}H_4]^{4-}$ were revisited by other researchers *(39,40)*. Even though early computational studies suggested a tetrahydroxy configuration for the $[\gamma\text{-}(SiO_4)W_{10}O_{32}H_4]^{4-}$ species (Scheme 5, species A) *(30,35)*, a combined computational and experimental analysis by Sartorel, Bonchio and coworkers supported an asymmetric protonation pattern, combining two terminal oxo groups and two aqua ligands in an antisymmetric fashion (Scheme 5, species B) *(39)*. Comparison of the relative stabilities of isomers A and B resulted in rather small energy differences between them, laying within a range of 6 kcal/mol. Still, the calculated structure of the bisaquo complex was found to be in better explain the distortions inducing chirality observed in the X-ray structure. This analysis was further extended to POM models with the general formula $[\gamma\text{-}(X^{n+}O_4)$

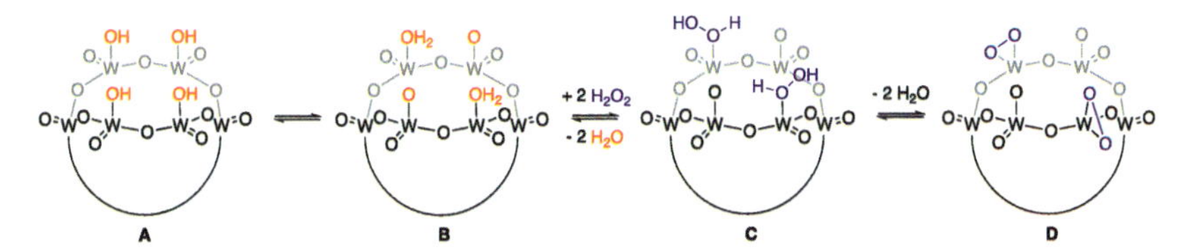

Scheme 5 Proposed acid/base equilibria involving terminal oxygen sites of the dilacunary γ-Keggin anion and their interaction with H_2O_2.

$W_{10}O_{32}H_4]^{(2-n)-}$ (X^{n+} = Si(IV), Ge(VI), P(V)) *(40)*, revealing in all cases a thermodynamic preference for the bis-aquo bis-oxo structure compared to the tetrahydroxy one. These observations prompted the authors to postulate that these aqua ligands may serve as leaving groups to allow the coordination of two H_2O_2 molecules, further evolving towards a bis-η^2-peroxo species upon dehydration (Scheme 5, C–D) *(40)*.

It is worth pointing out that regardless the proposed O-transfer pathway, computational studies on both systems consistently concluded that, despite the presence of multiple W sites within the metaloxo clusters, these act as single-site catalysts. This implies that both the heterolytic activation of the oxidant and the electrophilic oxygen transfer to the alkene take place on a single W(VI) ion.

Sartorel, Bonchio and coworkers also performed DFT calculations to gain insights into the structure-activity relationships governing the epoxidation reaction by lacunary Keggin-type polyoxotungstates *(40)*. For the [γ-($X^{n+}O_4$) $W_{10}O_{32}H_4]^{(2-n)-}$ (X^{n+} = Si(IV), Ge(VI), P(V)) structure the authors carried out a molecular orbital (MO) analysis on the bis-η^2-peroxo intermediate (Scheme 5, species D), which was assumed to act as the epoxidizing species. This revealed that the $\sigma^*_{O\text{–}O}$ MO of the peroxo group, which receives the attack from the $\pi_{C=C}$ MO of the alkene, is lower in energy for X= Ge and P compared to Si, thus predicting [γ-($X^{n+}O_4$)$W_{10}O_{32}H_4]^{(2-n)-}$ (X^{n+} = Ge(VI) and P(V)) as promising epoxidation catalysts *(40)*. Mizuno and coworkers followed a similar strategy with dinuclar peroxotungstates, exploring the role of the heteroatom in the XO_4^{n-} unit that connects both W(VI) centers *(36)*. The B3LYP-optimized geometries of $[X^{m+}O_4\{WO(O_2)_2\}_2]^{(4-m)-}$ (X= Se(VI), As(V), P(V), S(VI) and Si(IV)) complexes revealed that the epoxidation rate correlates with the W—OX distance; that is, the longer the distance (weakly coordinated XO_4^{n-} ligand), the faster the oxidation. This was ascribed to an increased Lewis acidity of the catalytically active W(VI) center, which decreases the charge density of peroxo groups, easing the rate-determining electrophilic O-transfer step. Thus, this infers that the reactivity of epoxidation catalysts can be modified by tuning of the electronic properties of the catalytic sites, opening avenues to the development of catalyst-design strategies.

More recently, Sing, Villanneau and coworkers synthesized a mono-protonated hybrid organic-inorganic B,α-$[NaHAsW_9O_{33}\{P(O)CH_2CH_2CO_2H\}_2]^{3-}$ POM, which is based on the arsenotungstate B,α-$\{AsW_9O_{33}\}^{9-}$ framework functionalized with two organophosphonyl groups *(41)*. This compound was successfully applied to the catalytic epoxidation of alkenes with H_2O_2. In this work, DFT calculations were carried out to identify the most likely positions of the proton within the POM structure and to investigate the formation of active epoxidizing species upon interaction with H_2O_2. These calculations indicate that oxygen atoms delimiting the POM lacuna are the most basic sites and hence, that one of the terminal {W=O} groups should be better formulated as {W-OH}. The interaction of the aforementioned model with H_2O_2 was found to yield a species bearing a W-OOH group in an almost ergoneutral process. This recalls to the early studies by Musaev and coworkers, which also proposed the formation of a W-OOH active species on the lacunary Keggin-type silicodecatungstate $[\gamma\text{-}(SiO_4)W_{10}O_{32}H_4]^{4-}$ *(30)*. However, the authors also analyzed the possible formation of other active species, finding that both a W-OO intermediate with a proton on a neighboring terminal oxygen and a peracid function on the carboxylic acid arm are accessible as well, being less stable than the W-OOH species by only less than 2.9 and 1.8 kcal/mol, respectively. Thus, the authors did not rule out that the electrophilic O-transfer to the alkene could take place from any of the characterized intermediates *(41)*.

3. Divanadium(V)-substituted γ-Keggin POMs

One of the first examples of a TM-substituted POM acting as an efficient and selective epoxidation catalyst with H_2O_2 consists in a divanadium-substituted γ-Keggin anion of $[\gamma\text{-}1,2\text{-}H_2SiV_2W_{10}O_{40}]^{4-}$ formula, which was reported by Mizuno and coworkers *(42)*. As shown in Fig. 4A, this consists in a dilacunary γ-Keggin-type silicotungstate, capped by a $\{OV\text{-}(\mu\text{-}OH)_2\text{-}VO\}$ core. Besides being highly active and chemoselective, thus providing high oxidation utilization efficiency, this system was found to grant unique stereospecificity, diastereoselectivity, and regio-selectivity compared to the vanadium-free $[\gamma\text{-}SiW_{10}O_{34}(H_2O)_2]^{4-}$.

Nakagawa and Mizuno performed DFT calculations at the B3LYP level in combination with experimental kinetic studies to shed light into the

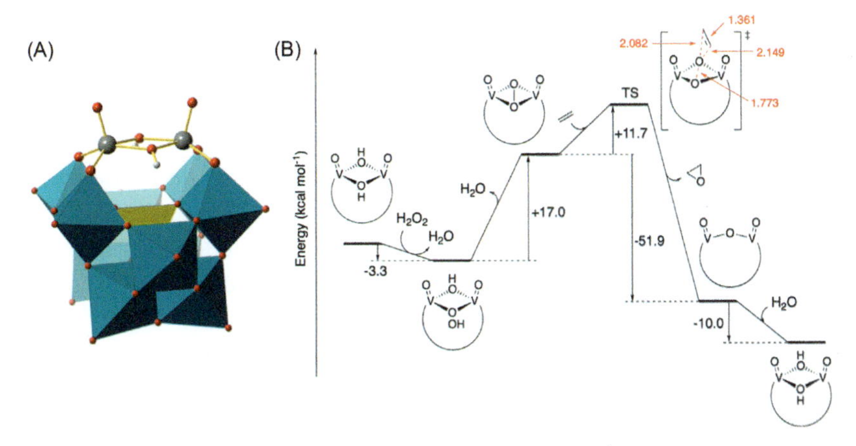

Fig. 4 (A) Molecular structure of the $[\gamma\text{-}1,2\text{-}H_2SiV_2W_{10}O_{40}]^{4-}$ catalyst. Vanadium(V) centers are shown as gray spheres, whereas W(VI) and Si(IV)-oxo groups are displayed as cyan and yellow octahedra, respectively. (B) Energy profile for the epoxidation of ethene by $[\gamma\text{-}1,2\text{-}H_2SiV_2W_{10}O_{40}]^{4-}$. Energy values (kcal/mol) and key distances (Å) of the TS structure are obtained from ref. *(43)*.

mechanism responsible for this catalytic process *(43)*. They suggested that upon interaction with H_2O_2, one of the bridging hydroxyl groups from the $\{OV\text{-}(\mu\text{-}OH)_2\text{-}VO\}$ core turns into a μ-hydroperoxo ligand, which further evolves to eliminate the second hydroxyl as a water molecule, yielding a species that contains a $\mu\text{-}\eta^2{:}\eta^2$-peroxo group. The latter was proposed to be the active species from which the electrophilic oxygen transfer to the alkene takes place, leveraging the cooperative effect of both vanadium atoms in activating the O–O bond (see Fig. 4B). Still, the formation of the $\mu\text{-}\eta^2{:}\eta^2$-peroxo species from the μ-hydroperoxo/μ-hydroxo intermediate was found to proceed uphill, with an energy cost of about 17 kcal/mol. Finally, the formation of the epoxide product was found to be strongly exothermic, providing the driving force for the reaction and rendering it irreversible (Fig. 4B). Additional DFT calculations were carried out to extend the analysis of this mechanistic proposal to several alkene substrates. These suggested that steric repulsion between the catalyst and the substituents of the alkene has an important impact on the activity and diastereoselectivity of the reaction *(43)*. In particular, steric effects explain why the epoxidation of the more accessible *cis* alkenes is easier than that of their *trans* isomer, despite the fact that they exhibit similar electronic properties.

The mechanism underlying the above reaction was then revisited by Musaev and coworkers *(44)*. This study also explored the role that solvent effects on the reaction through the IEF-PCM implicit solvent model and the incorporation of explicit water molecules; as well as that of explicit countercations. Overall, these calculations supported that the reaction proceeds via a peroxo pathway, as the hydroperoxo path was found to be more energy-demanding and hence, unlikely to compete with the former. Still, the authors showed that the polarization of the catalyst granted by the presence of the solvent, countercations and hydrogen-bonded water molecules interacting with the terminal V=O groups of the catalyst facilitate the overall reaction by reducing the height of the energy barrier *(44)*.

A few years later, Mizuno and coworkers tested at the experimental level the catalytic activity of the related $[\gamma\text{-}1,2\text{-}H_2PV_2W_{10}O_{40}]^{3-}$ system *(45)*, in which the silicon heteroatom of the previously studied $[\gamma\text{-}1,2\text{-}H_2SiV_2W_{10}O_{40}]^{4-}$ catalyst (Fig. 4A) is replaced by phosphorus, thus decreasing its overall negative charge from −4 to −3. This catalyst was found to outperform its silicon-containing analogue, in line with the computational predictions made by Sartorel, Bonchio and coworkers when comparing Vanadium-free $[\gamma\text{-}XW_{10}O_{34}(H_2O)_2]^{n-}$ POMs with $X = P(V)$ versus Si(IV) *(40)*. The authors also performed quantum chemical calculations at the HF/6–311 G(d,p) level on a series of C_6 alkenes and found that the energies of their $\pi_{C=C}$ orbitals do not correlate with the experimental reactivity trend. This was ascribed to steric constraints between hindered alkenes and the catalyst, further supporting previous observations by Nakagawa and Mizuno on $[\gamma\text{-}1,2\text{-}H_2SiV_2W_{10}O_{40}]^{4-}$-catalyzed epoxidation processes *(43)*. As a matter of fact, whereas the $\pi_{C=C}$ MO energy of *cis*- and *trans*-2-butene were found to be nearly identical (−6.631 vs −6.623 eV respectively), DFT calculations revealed a lower barrier for the electrophilic oxygen transfer from the $\mu\text{-}\eta^2{:}\eta^2$-peroxo species to the *cis* isomer (*ca.* 12.2 kcal/mol) compared to the *trans* one (*ca.* 15.3 kcal/mol) *(45)*. This confirmed that the C=C double bonds in *cis* alkenes are indeed more accessible and easier to epoxidize.

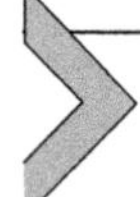

4. Learnings from single-site Titanium(IV)-containing POMs

4.1 Characterization of active species and reaction mechanism

Ti(IV)-substituted POMs constitute the broadest family of POM-based epoxidation catalysts with H_2O_2. As such, they have been significantly more explored from both computational and experimental perspectives than other TM-substituted POMs, including the above-described V-containing POMs. This can be explained by the fact that Ti-POMs have been extensively used as tractable molecular models of the heterogeneous Titanium-silicalite (TS-1) catalyst that is used at the industrial scale, providing a robust and well-defined coordination sites for Ti(IV) ions. Even though Ti-POMs were originally designed to gain mechanistic insights into the epoxidation of alkenes with H_2O_2 catalyzed by TS-1, they proved highly active and selective catalysts per se, sometimes with activities that compare to those of their heterogeneous counterpart *(1,25)*.

Early computational studies on Ti-POMs focused on the characterization of the active Ti-peroxo complex $[HPTi(O_2)W_{11}O_{39}]^{4-}$ *(46)*, which was experimentally found to form upon hydrolysis of the μ-hydroxo dimeric heteropolytungstate $[(PTiW_{11}O_{39})_2OH]^{7-}$ in acetonitrile in the presence of an excess of H_2O_2 *(47)*. DFT calculations determined that the most stable molecular configuration for $[HPTi(O_2)W_{11}O_{39}]^{4-}$ corresponds to a species with a terminal η^2-peroxo group and a protonated Ti−O−W site (Fig. 5A). However, the small energy difference with the η^1-hydroperoxo species suggested that both Ti−OH−W and TiOO−H protonated anions may coexist in solution.

A subsequent work aimed at the more ambitious goal of characterizing of the reaction mechanism responsible for the catalytic epoxidation activity of Ti (IV)-POMs *(48)*. These mechanistic studies focused on two types of catalysts, both represented in Fig. 6: the Ti-monosubstituted Keggin-type POM $[PTi(OH)W_{11}O_{39}]^{4-}$ introduced above and on the Ti-disubstituted sandwich-type dititanium 19-tungstodiarsenate(III) POM $[Ti_2(OH)_2As_2W_{19}O_{67}(H_2O)]^{8-}$, which had been also applied to the catalytic epoxidation of alkenes *(34)*. In the former, the $[PW_{11}O_{39}]^{7-}$ lacunary species acts as a pentadentate ligand for the hexa-coordinated Ti(IV) ion; whereas in the sandwich-type POM, two $\{B\text{-}\alpha\text{-}As^{III}W_9O_{33}\}$ Keggin moieties clamp two penta-coordinated, square-pyramidal $[Ti(OH)]^{3+}$ groups and an octahedral $[WO(H_2O)]^{4+}$ fragment.

While the prevalence of a peroxo path was well established that for divanadium-containing POMs *(43–45)* and generally accepted for Ti(IV)-based catalysts, computational studies on model titanosilicalites had shown that an O-transfer mechanism involving the Ti-OOH species proceeds through a lower energy barrier *(49,50)*. Thus, this study fully characterized both hydroperoxo and peroxo paths for the analyzed Ti-POM catalysts.

Following the heterolytic activation of the H_2O_2 oxidant on a Ti-OH site and an intramolecular proton transfer from the as formed TiOOH group

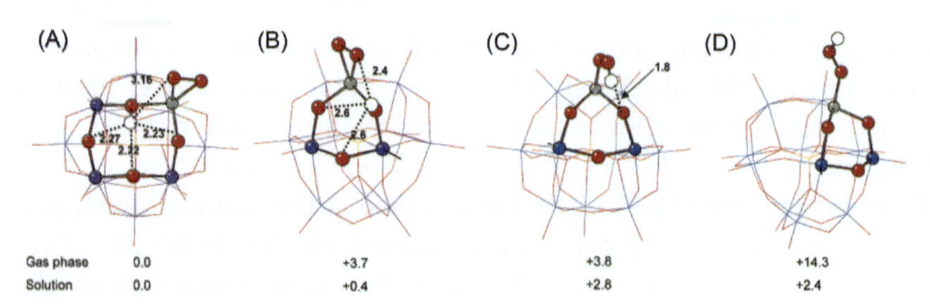

Fig. 5 Relative energies (kcal/mol) of several structural isomers of $[HPTi(O_2)W_{11}O_{39}]^{4-}$ calculated at the BP86/triple-ζ+polarization level. Energies of solvated anions (bottom row) were obtained through the incorporation of solvent effects of acetonitrile via the COSMO implicit model. Relevant distances are given in Å. *Reprinted with permission from Inorg. Chem.* ***2004****, 43, 2284–2292 (46). Copyright 2024 American Chemical Society.*

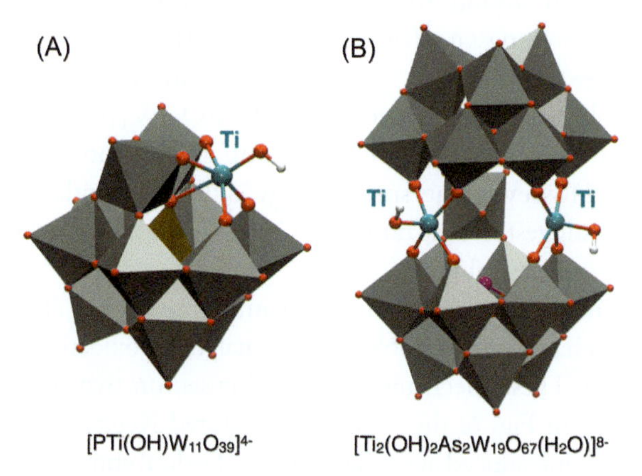

Fig. 6 Combined balls-and-sticks and polyhedral representation of Ti(IV)-containing $[PTi(OH)W_{11}O_{39}]^{4-}$ (A) and $[Ti_2(OH)_2As_2W_{19}O_{67}(H_2O)]^{8-}$ (B) catalysts.

to a neighboring bridging oxygen site (see TS1A and TSA' in Fig. 7), which take place through moderate energy barriers, both POMs were found to yield a peroxo TiOO intermediate protonated at a Ti-O-W site *(48)*. For both POMs in Fig. 6, the TiOO intermediate was found to be slightly more stable than the TiOOH one, concluding that the Ti-peroxo species is likely acting as the resting state of the catalyst. However, the analysis of the electrophilic O-transfer step to the alkene revealed that the hydroperoxo path, whereby the Ti-OOH species acts as the real epoxidizing species, is more energetically favored (see Fig. 7, black vs red-dashed lines). This was explained by the stronger polarization of the O–O bond granted by the presence of a proton in the hydroperoxo group, which is even more

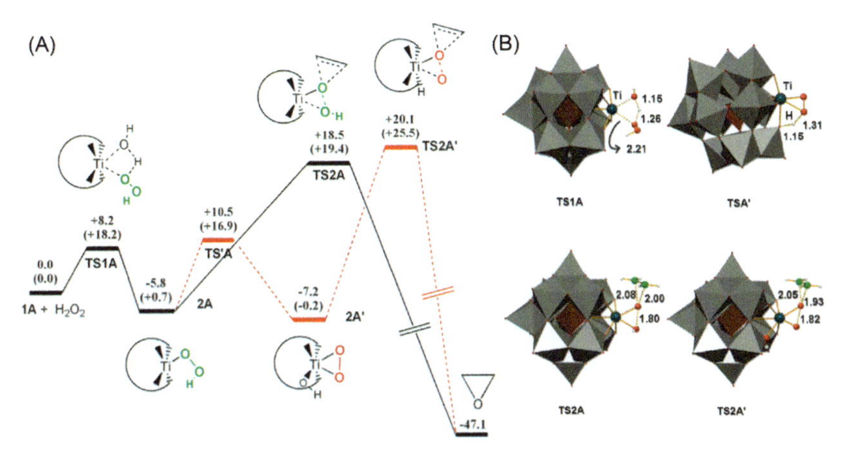

Fig. 7 (A) Electronic energy profile (kcal/mol) for the epoxidation of ethene by $[PTi(OH)W_{11}O_{39}]^{4-}$, labeled as 1A, obtained with the B3LYP exchange-correlation functional and double-ζ+polarization-quality basis sets. Values in parentheses stand for energies obtained in the presence of solvent effects of acetonitrile, included through the SMD solvent model. (B) Transition state structures involved in the reaction mechanism, in which key distances are given in Å. *Adapted with permission from J. Am. Chem. Soc.* ***2010****, 132, 7488–7497 (48). Copyright 2024 American Chemical Society.*

noticeable in the presence of a polar solvent such as acetonitrile *(48)*. In fact, this protonation results in a polarization of the $\sigma^*_{O\text{-}O}$ orbital towards the non-protonated O_α of the hydroperoxo ligand and therefore, the attack of the alkene was considered to take place on this site. Remarkably, this differs from the initial mechanistic proposal by Musaev and co-workers based on Keggin-type silicodecatungstate $[\gamma\text{-}(SiO_4)W_{10}O_{32}H_4]^{4-}$, in which the protonated O_β of a W-OOH intermediate was proposed to be transferred to the double bond of the alkene, being assisted by a water molecule acting as a proton shuttle *(30,35)*.

The calculated energy barriers for the sandwich dimer represented in Fig. 6B were found to be higher than those obtained for the Ti-mono-substituted Keggin catalyst, concluding that the lower coordination number for Ti(IV) sites of the former (5- vs 6-fold) is not responsible for its experimentally observed higher activity *(34)*. Instead, this reactivity trend was ascribed to the fact that the steric crowding around Ti centers provided by tungstate groups of the catalyst prevents its dimerization into inactive species, having a positive impact on the activity *(48)*. Moreover, it is worth stressing that despite the presence of two Ti(IV) ions in the catalyst structure, this was found to operate as a single-site catalyst. This is due to the fact that both Ti centers are too far apart to enable the type of cooperative mechanisms characterized for divanadium-containing γ-Keggin catalysts (*vide supra*) or for the analogous dititanium-containing $[\gamma\text{-}1,2\text{-}H_2SiTi_2W_{10}O_{40}]^{4-}$ anions, applied to the catalytic oxidation of thioethers *(51)*.

For the "sandwich"-type dititanium POM discussed above, DFT calculations were also conducted to investigate the correlation between the catalytic activity and the properties of the alkene substrates *(52)*. These studies concluded that the activity is primarily governed by the nucleophilicity of the alkene, described by the energy of the $\pi_{C=C}$ orbital as obtained from NBO analysis. Still, the performance of QM/MM calculations using the ONIOM method revealed that steric interactions play a secondary role in specific cases, such as in the epoxidation of *trans*-stilbene, which experimentally proved less active than its *cis* isomer. Specifically, whereas the *cis*-stilbene accommodates both phenyl substituents far away from the catalyst in the O-transfer TS geometry, in the *trans* isomer one phenyl substituent has to face the POM structure, thereby inducing steric repulsion and slightly hampering its epoxidation *(52)*. Similar conclusions were previously drawn by Nakagawa and Mizuno while analyzing the reactivity of a divanadium-containing POM (*vide supra*) *(43)*.

4.2 Impact of the protonation state

Experimental studies on the Ti-monosubstituted Keggin POM by Kholdeeva and coworkers evidenced that the protonation state of the active Ti-(hydro) peroxo species plays a key role in its catalytic activity *(53)*. Specifically, the activity was found to rise as the number of protons in the Ti-POM structure increases. This effect was thoroughly investigated in a combined computational and experimental work *(54)*. Specifically, this study compared the performance of the monoprotonated $[HPTiW_{11}O_{40}]^{4-}$ with that of the deprotonated $[H_2PTiW_{11}O_{40}]^{3-}$, formed upon addition of one equivalent of acid to a solution of the monoprotonated anion in acetonitrile.

While the position of the first proton in $[HPTi(O_2)W_{11}O_{39}]^{4-}$ is widely accepted to correspond to a Ti-O-W site, that of the second one had never been investigated before. Therefore, DFT calculations were initially conducted to identify the most likely position for the second proton in $[H_2PTi(O_2)W_{11}O_{39}]^{3-}$ *(54)*. A systematic exploration of possible isomers indicated that an intermediate with two protons at Ti-O-W sites is more stable by 3.4 kcal/mol in acetonitrile than a Ti-hydroperoxo species with only one proton located in a Ti-O-W site. Still such a small energy difference did not allow ruling out that both species coexist.

Further calculations were carried out to investigate the impact of the presence of a second proton on the ability of the catalyst to promote alkene epoxidation using ethene as a model substrate *(54)*. These calculations indicate that the hydroperoxo path is preferred over the peroxo one regardless the protonation state of the Ti-POM. Most importantly, the calculated energy barrier for the O-transfer step from the diprotonated catalyst was found to be significantly lower than that occurring from the monoprotonated one. In fact, the barrier was found to drop by *ca.* 6 kcal/mol, in excellent agreement with experimentally-derived Arrhenius activation energies. Moreover, similar results were obtained when varying the position of the second proton, pointing towards a *global charge* effect, whereby the decrease of the negative charge of the POM induced by protonation renders the Ti-OOH group more electrophilic and in turn, more prone to transfer an oxygen atom to the nucleophilic double bond of the alkene.

A few years later, DFT calculations on Nb(V)-monosubstituted POMs (detailed below) further supported these conclusions *(55)*. Moreover, this work established a comparison between Lindqvist and Keggin structures in terms of protonation effect. While the Lindqvist anion was found to be highly efficient for alkene epoxidation, the Keggin-based analogue displayed significantly poorer performance. In this case, DFT calculations indicate that the Keggin anion is not basic enough to associate protons and consequently, to grant efficient activation of H_2O_2 and subsequent alkene epoxidation *(55)*. Overall, this infers that the POM scaffold needs to be negatively charged enough to enable the association of protons, but not too much to preserve the electrophilic character of the oxidizing species required for the electrophilic O-transfer step.

4.3 Impact of the POM structure

DFT calculations have also provided unique insights into how the structural features of the POM support influence the catalytic activity of embedded metal centers. As introduced in the previous section, protonation was found to affect the reactivity through a *global charge* effect, which translates into variations in the electrophilicity of the active Ti-OOH group. Hence, it is reasonable to think that the charge density underlying POM scaffolds can also affect the electronic properties of catalytic sites in a similar fashion. In fact, early studies discussed above had also evidenced that the catalytic activity of Keggin-based catalysts can be improved by increasing the positive charge of the central heteroatom, thus lowering the overall negative charge of the POM *(40,45)*. Additionally, the POM scaffold can be also designed to tune the topology of binding sites for active metal centers, leading to a powerful strategy to control their coordination environment. As detailed in the following subsections, this has crucial implications as well, not only from a mechanistic standpoint, but also in terms of product selectivity.

4.3.1 Rigidity imposed by full-inorganic POM ligands: O_α versus O_β transfer

After initial calculations, Kholdeeva and co-workers determined the Arrhenius activation energy (E_a) for cyclooctene epoxidation in the presence of $TBA_{5.5}Na_{1.55}K_{0.5}H_{0.5}[Ti_2(OH)_2As_2W_{19}O_{67}(H_2O)]$ *(1)*, which turned out to be lower than the calculated energy barrier for the $[Ti_2(OH)_2As_2W_{19}O_{67}(H_2O)]^{8-}$ catalyst operating through an hydroperoxo path whereby the non-protonated O_α is transferred to the alkene *(48)*. In the context of thioether oxidation, DFT calculations showed that for the tetratitnanium-containing POM $[\{\gamma\text{-}SiTi_2W_{10}O_{36}(OH)_2\}_2(\mu\text{-}O)_2]^{8-}$, sulfoxidation is preferred through a O_β-transfer from the Ti-OOH species group because the O_α-transfer involves an unfavorable 7-fold coordinated Ti environment *(51)*. These findings prompted revisiting the mechanism underlying the alkene epoxidation by $[Ti_2(OH)_2As_2W_{19}O_{67}(H_2O)]^{8-}$, in which, given the sterically hindered nature of the Ti-OOH groups formed upon interaction with H_2O_2, transferring the more accessible O_β center may be also preferred *(56)*.

Indeed, the O_β-transfer was found to be preferred by more than 6 kcal/mol over the O_α-transfer mechanism for the dititanium-containing sandwich-type POM, as illustrated in Fig. 8. Note that in addition to such a mechanistic switch, protonation of the catalyst was identified to be essential to reproduce the experimental activation energy, suggesting that the active species should contain a proton besides that of the hydroperoxo group. Conversely, the Ti(IV)-monosubstituted Keggin catalyst, in which the Ti-OOH group is found in a more flexible and less sterically hindered environment, was confirmed to operate preferentially by transferring the more electrophilic O_α center *(56)*.

The above conclusions were also found to extend to the explanation of the reactivity of the Ti(IV)-monosubstituted Lindqvist anion $[Ti(OH)W_5O_{18}]^{3-}$ *(57)*. Unlike in the Keggin structure, the Ti(IV) center embedded in the Lindqvist structure is found in a stiffer coordination environment imposed by the molecular arrangement of the five chelating oxygen centers of the POM (see Fig. 9). As a consequence, for the Ti-Lindqvist catalyst, the

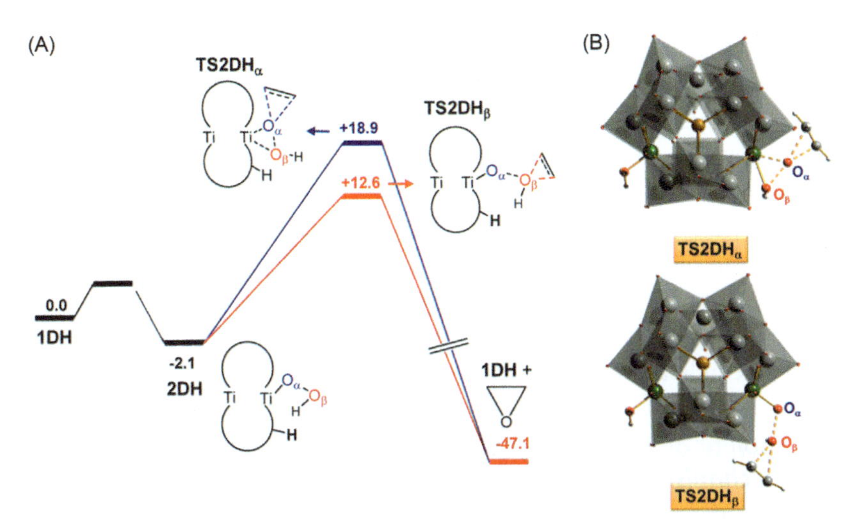

Fig. 8 Comparison of the electronic energy profiles (kcal/mol) for the epoxidation of ethene by following O_α- and O_β-transfer paths. *Adapted with permission from Inorg. Chem.* ***2016****, 55, 6080–6084 (56). Copyright 2024 American Chemical Society.*

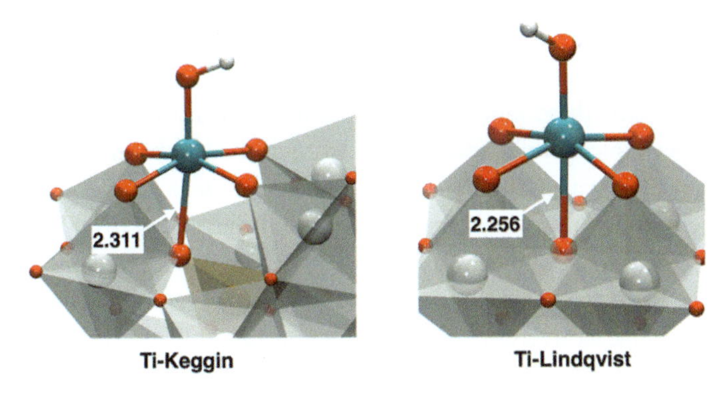

Fig. 9 Comparison of the Ti(IV) coordination environment when embedded in a Keggin (left) or a Lindqvist (right) POM structure. Ti—O distances for the B3LYP-optimized structures are given in Å.

β-oxygen transfer pathway becomes energetically favored over the α-oxygen hydroperoxo (by ca. 3 kcal/mol in terms of Gibbs free energy for the epoxidation of cyclohexene). This indicates that the rigid environment of the Ti center makes it reluctant to increase its coordination number from six to seven to accommodate the O_α-transfer TS and thus, the penalty for transferring the less electrophilic β-oxygen is lower than that of reaching a distorted TS configuration *(57)*. It is also worth noting that mechanistic investigations on the $[Ti(OH)W_5O_{18}]^{3-}$ catalyst revisited the H-transfer steps responsible for the Ti-OH/Ti-OOH and TiOOH/Ti(OO)-O(H)-W equilibria, finding that the assistance of an external water molecule acting as a proton shuttle lowers the free-energy barriers for these processes by ca. 4 and 9 kcal/mol, respectively *(57)*. Moreover, this work reported that upon the incorporation of entropic effects, the Ti-η^2-(OO)H species is slightly more thermodynamically stable than the previously characterized Ti-η^1-(O)OH configuration, strongly suggesting that both coordination modes might coexist in equilibrium, at least for the Ti-Lindqvist catalyst *(57)*.

4.3.2 Lability of Ti–O bonds: inner- versus outer-sphere O-transfer mechanism

Leveraging the expertise of the group of A. Proust in the synthesis of hybrid organic-inorganic POMs, Guillemot and coworkers developed a $[SbW_9O_{33}(^tBuSiOH)_3]^{3-}$ silanol-decorated polyoxotungstate, which offers a tripodal coordination site for Ti(IV), leading to the $[\alpha\text{-B-}SbW_9O_{33}(^tBuSiO)_3Ti(O^iPr)]^{3-}$ complex, represented in Fig. 10 *(58)*. Notably, this complex fairly mimics the coordination environment of Ti centers in the heterogeneous TS-1 epoxidation catalyst, being relevant given that the catalytic performance of TS-1 has been largely attributed to the environment of Ti(IV) embedded in the silica lattice *(59)*. In fact, these kinds of hybrid silanol-lacunary POM structures had previously been successfully used to model vicinal silanols on dehydroxilated silica *(60)* and low-valent V(III) intermediate single-atom silica-supported catalysts, which are involved in Mars–van Krevelen oxidations *(61)*. Experimentally, the $[\alpha\text{-B-}SbW_9O_{33}(^tBuSiO)_3Ti(O^iPr)]^{3-}$ complex showed the ability to selectively epoxidize allylic alcohols at room temperature, achieving high conversions, while conversions for nonfunctionalized alkenes under the same conditions were reported to remain insignificant *(58)*.

Poblet, Guillemot and coworkers carried out a thorough computational investigation by means of DFT calculations aimed at understanding the reaction mechanism at play as well as the origin of the observed selectivity towards allylic alcohols *(62)*. This study revealed that the epoxidation reaction of 1-methyl-2-buten-1-ol occurring through conventional *outer-sphere* hydroperoxo pathways (Fig. 11, left) involves too high free-energy barriers (>27 kcal/mol) to reproduce the value of 22.3 kcal/mol derived from the experimental rate constant. In this case, the Ti-peroxo complex protonated at a neighboring Ti-O-Si site was found to be significantly less stable than the Ti-OOH species, leading to prohibitively high barriers as well.

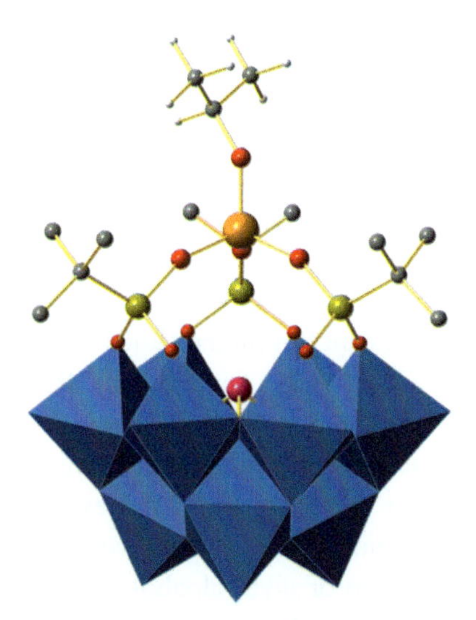

Fig. 10 Combined balls-and-sticks and polyhedral representation of the hybrid organic-inorganic $[\alpha\text{-B-SbW}_9O_{33}(^t\text{BuSiO})_3\text{Ti}(O^i\text{Pr})]^{3-}$ POM catalyst. H atoms in ^{t}Bu groups are omitted for clarity. Color code: Ti (orange), Si (yellow), W (blue), Sb (magenta), O (red), C (gray), H (white). *Reprinted with permission from ACS Catal.* ***2020****, 10, 4737–4750 (62). Copyright 2024 American Chemical Society.*

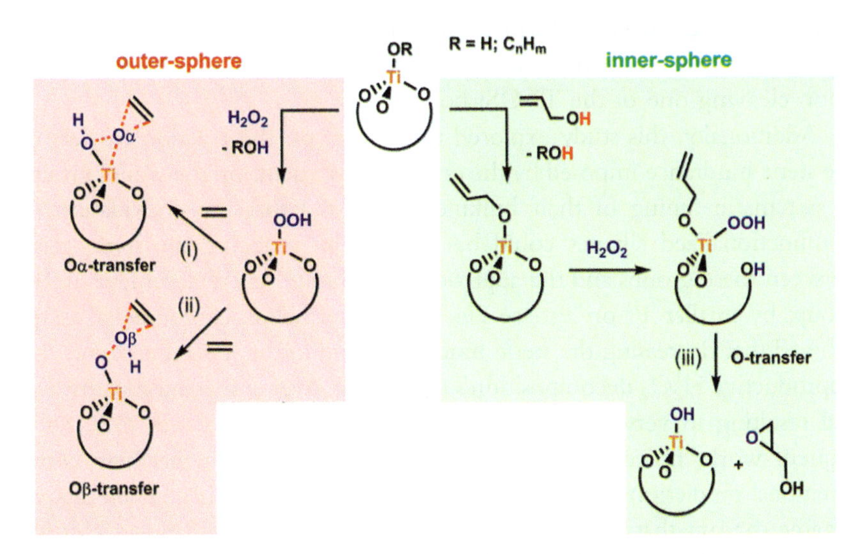

Fig. 11 Possible mechanisms for alkene epoxidation with H_2O_2 catalyzed by Ti-containing silanol-decorated polyoxotungstates.

Nevertheless, the authors characterized a novel *inner-sphere* reaction mechanism (Fig. 11, right), unprecedented for Ti-POMs that involves the coordination of both the oxidant and the substrate to the active Ti(IV) center. This takes place through the formation of a Ti-alcoxy resting state formed upon interaction via protolysis of the allylic alcohol substrate and the catalyst precursor bearing either a hydroxo or another alcoxy (^{i}PrO) group. The formation of this resting state was supported by NMR measurements *(58)*. Subsequently, the H_2O_2 oxidant can interact with the catalyst via protolysis of a labile Ti-OSi junction. This yields a flexible Ti-(OR)(η^2-OOH) (R = allyl) metastable intermediate (species 5 in Fig. 12), from which the epoxidation of the double bond takes place through a calculated overall free-energy barrier of 23.0 kcal/mol, in excellent agreement with the experimental value of 22.3 *(62)*. Hence, besides providing atomically-resolved details about the experimentally observed epoxidation of allylic alcohols, these results clearly indicate that the reported inertness of nonfunctionalized alkenes can be attributed to the fact that they cannot bind to the Ti(IV) center to react through the *inner-sphere* mechanism. Instead, their epoxidation can only be attained through more energy demanding *outer-sphere* pathways.

The structure of the TS for the novel inner-sphere path is shown in Fig. 12, and compared to that of the conventional outer-sphere mechanism. As illustrated in Fig. 12, the reduction of the free-energy barrier observed in going from the outer- to the inner-sphere O-transfer mechanism was ascribed to an increased flexibility of the Ti(IV) center upon cleaving one of the Ti-OSi bonds.

Additionally, this study explored the impact on the reaction barriers of the steric hindrance imposed by the organic substituents on the silanol group. A systematic tuning of their bulkiness suggested that the epoxidation of nonfunctionalized alkenes could be attained by reducing the steric clash between these groups and the approaching substrate, via replacement of ^{t}Bu groups by smaller ^{i}Pr or ^{n}Pr (see Fig. 13). Nevertheless, experimental assays showed that decreasing the steric hindrance around the Ti center favors the unproductive H_2O_2 decomposition side reaction over epoxidation pathways, still resulting in very low yields for nonfunctionalized alkenes *(62)*. Subsequent work, however, provided experimental evidence for these computational predictions using tertbutyl hydroperoxide (TBHP), an alternative organic oxidant that is less sensitive to decomposition, increasing yields for cyclohexane epoxidation from $< 5\%$ in the ^{t}Bu-containing system to 70% and 97% with ^{i}Pr or ^{n}Pr substituents, respectively *(63)*.

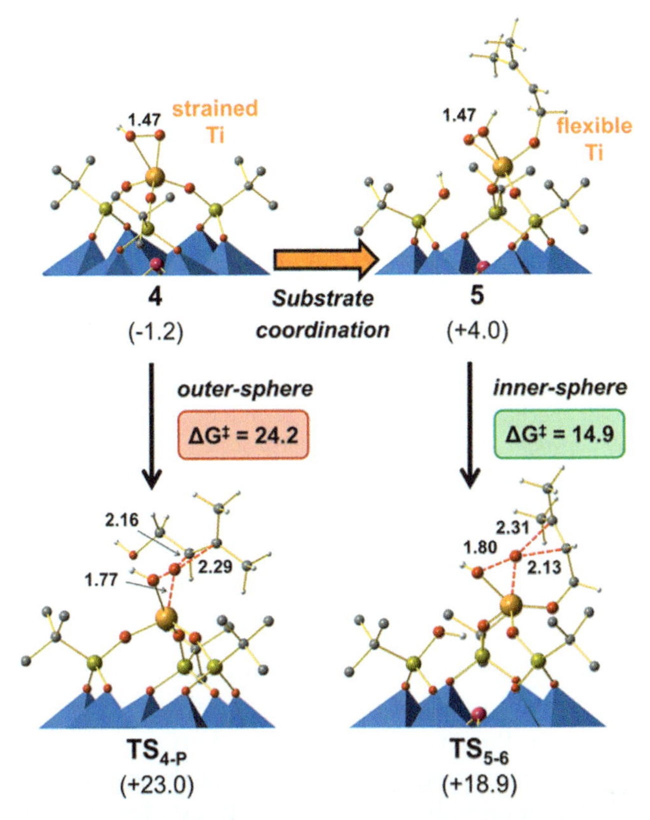

Fig. 12 Comparison of the calculated free-energy barriers (kcal/mol) for allylic alcohol epoxidation by the Ti-containing hybrid organic-inorganic POM through outer- (left) and inner-sphere (right) pathways. Relative Gibbs free energies are given in parentheses with respect to the Ti-O^{i}Pr precursor. Red-dashed lines represent those bonds being formed or broken in the TS structure. Main distances are shown in Å. Hydrogen atoms of *t*Bu groups are omitted for clarity. *Reprinted with permission from ACS Catal.* ***2020****, 10, 4737–4750 (62). Copyright 2024 American Chemical Society.*

Overall, these studies highlight the importance of the structural features of the POM support in coordinating catalytically active metal centers for alkene epoxidation, demonstrating that tailored reactivity and selectivity can be achieved through the careful design and tuning of the metal center's environment.

5. Impact of the nature of the transition metal center

The fundamental evaluation of the role of the embedded metal center on the epoxidation reaction can become a challenging task at the

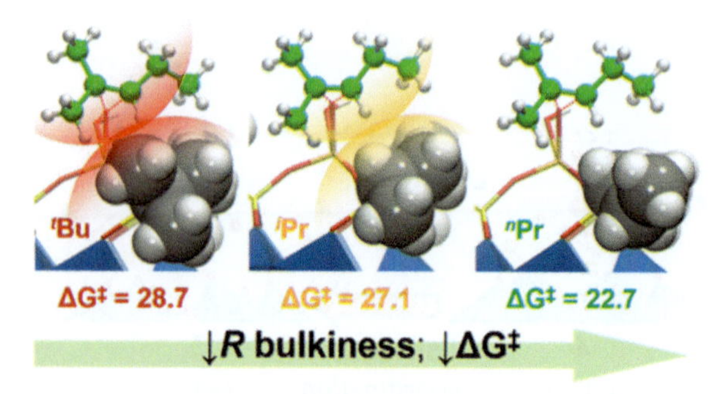

Fig. 13 Impact of the bulkiness of the organic substituents on the silanol function on the free-energy barrier for the epoxidation of a nonfunctionalized alkene via outer-sphere O-transfer mechanism. *Adapted with permission from ACS Catal.* ***2020****, 10, 4737–4750 (62) Copyright 2024 American Chemical Society.*

experimental level, mostly because the catalytic outcomes upon replacing metal centers may reflect other effects related to stability or supramolecular effects such as solubility, agglomeration or ion-pairing phenomena.

In this regard, computational chemistry allows for systematic variation of the metal center without affecting any other parameter, enabling the individual evaluation of the role of the metal center. Leveraging this strategy, Kholdeeva, Poblet, Carbó and coworkers assessed the impact of the nature of the substituting metal center by analyzing computationally a series of d^0 TM-monosubstituted Keggin POMs bearing a peroxo or hydroperoxo group $[PTM^{n+}(O_2)(H)W_{11}O_{39}]^{(8-n)-}$ (TM^{n+} = Ti^{IV}, V^V, Zr^{IV}, Nb^V, Mo^{VI}, W^{VI} and Re^{VII}) *(48)*. These calculations showed that the energy barriers for peroxo and hydroperoxo mechanisms follow opposite trends. In moving from the left to the right in the periodic table, the energy barrier for the peroxo path decreases significantly, while the energy barrier for hydroperoxo increases but to a lower extent (Fig. 14). Going down in the periodic table, both energy barriers decrease, this effect being greater for the hydroperoxo path. Overall, these results inferred that in moving from left to right and down in the periodic table, the O-transfer step through the peroxo path becomes competitive with the hydroperoxo path (see Fig. 14). This was attributed to a stronger mixing between the TM-centered orbitals and those of the peroxo group, which increases its electrophilic character and is less pronounced in the hydroperoxo mechanism *(48)*. These conclusions were later further supported by computational studies on a series of dimeric $[SeO_4WO(O_2)_2TMO(O_2)_2]^{n-}$ structures (TM = Ti^{IV}, V^V, Ta^V,

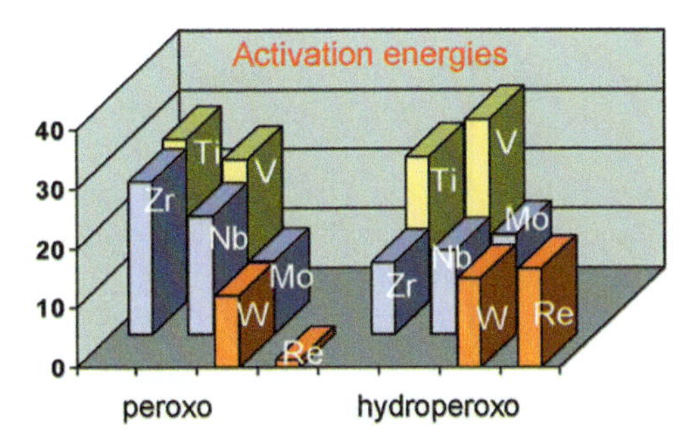

Fig. 14 Calculated activation energies (kcal/mol) associated with the oxygen transfer step to ethene via either the peroxo or hydroperoxo path for TM-substituted Keggin-type phosphotungstates in the gas phase. *Adapted with permission from J. Am. Chem. Soc.* ***2010****, 132, 7488–7497 (48). Copyright 2024 American Chemical Society.*

Mo^{VI}, W^{VI}, Tc^{VII}, and Re^{VII}) *(64)*, finding the same trends on the height of the O-transfer barriers that were observed for TM-monosubstituted Keggin anions.

In the case of W^{VI}-based catalysts, one should also note that even though rather low barriers (12–15 kcal/mol) were obtained for the non-substituted Keggin anion (TM = W^{VI}), this suffers from inability to activate H_2O_2, owing to the poor basicity of W=O sites in plenary polyoxotungstate structures that preclude the formation of terminal W-OH groups. This fact explains why lacunary tungstates with available hydroxyl groups are active catalysts for epoxidation reactions with H_2O_2 while plenary analogues with terminal oxo ligands are not *(48)*. It is also worth noting that the V(V)-monosubstituted Keggin was found to operate preferably through the peroxo path, in line with previous calculations on divanadium-substituted POMs *(43,44)*.

More recently, a series of combined experimental and theoretical works explored the impact of the nature of the metal center embedded in a Lindqvist structure, $[TM^{n+}(OH)W_5O_{18}]^{(7-n)-}$ (TM = Ti(IV), Nb(V), Zr (IV)) *(55,57,65)*. Experimentally, the catalytic activity at 50 °C in acetonitrile was found to increase in the order Ti(IV) < Nb(V) < Zr(IV). Interestingly, this trend aligns well with the computational predictions of the O-transfer barriers through the hydroperoxo path made on TM-Keggin POMs (*vide supra*). A complete mechanistic characterization of the possible reaction pathways for all three metals indicates that trend in the calculated, overall free-energy barriers for the most likely pathways can nicely reproduce the experimental reactivity trend (Fig. 15), further validating the mechanistic proposals. Furthermore, several conclusions can be drawn from this comparative analysis:

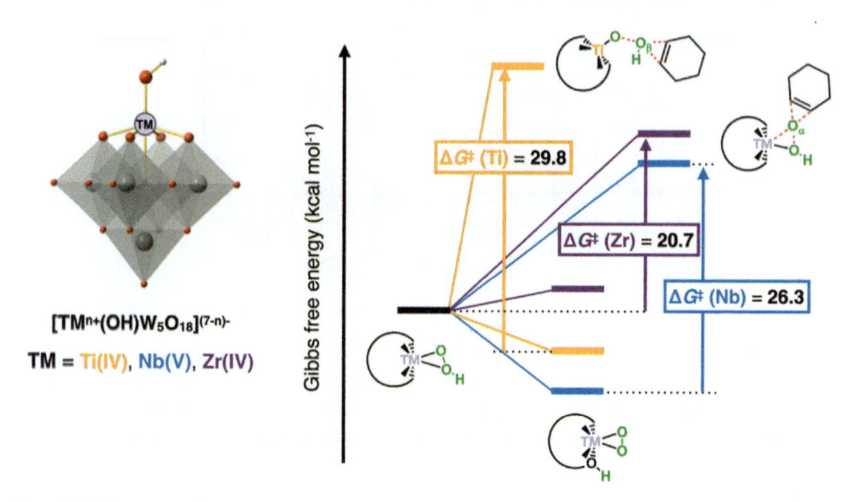

Fig. 15 Schematic free-energy profile comparing the catalytic performance of Ti(IV)-, Nb(V)- and Zr(IV)-monosubstituted Lindqvist anions towards the epoxidation of cyclohexene (orange, blue and purple lines, respectively). Overall free energy barriers are given for each catalyst in kcal/mol.

1. In going from the 4th to the 3rd row on the periodic table, the metal center becomes smaller and stiffer. This translates into the fact that unlike Nb(V) and Zr(IV) (4th row), which can efficiently accommodate distorted 7-fold coordinated O_α-transfer TSs, the Ti(IV) center (3rd row) is more likely to transfer the less electrophilic O_β center, explaining that it requires overcoming the highest free-energy barrier among the series.
2. In going from the 4th to the 5th period within the same row (Zr(IV) vs Nb(V)), the electrophilicity of the TM-OOH group increases due to a reduction of the overall negative charge of the POM. As a matter of fact, the O_α-transfer TS with respect to the TM-OOH intermediate lies lower in energy for Nb(V) than that of Zr(IV) (Fig. 15). However, the stronger Lewis acidity of Nb(V) compared to Zr(IV) is also responsible for the formation of a more stable peroxo resting state, leading to a higher overall free-energy barrier (Fig. 15).

6. Products selectivity and H_2O_2 decomposition side reaction

As already introduced in Section 1, the unproductive decomposition of H_2O_2 is the main side reaction that competes with the epoxidation of alkenes, being detrimental for both the selectivity of the reaction but also for the oxidant utilization efficiency. However, the lack of mechanistic knowledge over time has hindered the development of design rules to control or avoid it.

Originally, it was proposed that the decomposition of H_2O_2 over titanium-silicate catalysts involves the homolytic cleavage of the O–O bond in a Ti-OOH species, generating $TiO^{\bullet}$ and $^{\bullet}OH$ radicals that evolve to form degradation-related products *(66)*. However, DFT calculations by Don Tilley and co-workers revealed that such process is energetically inaccessible *(67)*. The authors rather proposed that the key step involves the interaction between the Ti-OOH intermediate and a second H_2O_2 molecule, resulting in the formation of $TiO^{\bullet}$ and $HO_2^{\bullet}$ radicals in addition to a water molecule. EPR studies confirmed the formation of superoxide radicals both adsorbed and covalently bound to Ti upon H_2O_2 decomposition over Ti-silicates *(68,69)*, although $TiO^{\bullet}$ species had never been identified by means of spectroscopic techniques. Even so, for years, the above mechanistic proposal has been adopted and used to explain selectivity trends in POM-catalyzed epoxidation reactions.

Kholdeeva, Poblet, Carbó and coworkers calculated the energy cost of the $^{\bullet}OH$ transfer to the alkene from the Ti-OOH intermediate as a measure of the ability of a Ti(IV)-Keggin POM to undergo a homolytic oxidation mechanism *(54)*. The goal was to give an explanation to the enhanced selectivity towards heterolytic products triggered by protonation. As detailed in Section 4.2, energy barriers for the heterolytic O-transfer step were found to decrease upon protonation. However, the energy cost for the homolytic O-O bond cleavage was found to be less sensitive to the protonation state. On these grounds, the authors proposed that the impact of protonation on the reaction selectivity mainly arises from the acceleration of the epoxidation path, making it more kinetically favorable over homolytic side processes *(54)*. A similar approach was adopted to explain the higher selectivity observed for a Nb(V)-monosubstituted Lindqvist compared to its Ti(IV)-based analogue *(57)*.

It was not until very recently, that a combined experimental and computational work investigated in detail the mechanism responsible for the H_2O_2 decomposition, taking the case of a Zr(IV)-monosubstituted Lindqvist anion as a representative example *(70)*. EPR measurements confirmed the formation of free $HO_2^{\bullet}$ and Zr-bound superoxo radical species, whereas catalytic experiments with test substrates supported the formation of singlet oxygen over the course of the H_2O_2 decomposition reaction.

DFT calculations allowed proposing the reaction mechanism summarized in Schemes 6 and 7. From the Zr-OOH species, a second H_2O_2 molecule can be heterolytically activated on the Zr center, forming a protonated Zr-dihydroperoxo intermediate. From the latter, the authors proposed an intramolecular nucleophilic attack of one hydroperoxo group to the other, forming an unprecedented Zr-trioxidate complex (Zr-OO(OH)) in addition to a water molecule (Scheme 6). The latter is formed in an overall exergonic process, overcoming an accessible free-energy barrier of 19.2 kcal/mol from the Zr(OOH)(O(H)OH) species. Importantly, as shown in Scheme 7, the Zr-trioxidate intermediate was found to be able to evolve either heterolytically, releasing singlet oxygen through a low barrier of 9.3 kcal/mol; or via homolytic ZrOO—OH bond cleavage, generating the experimentally observed radical species with an energy cost of only 8.3 kcal/mol *(70)*.

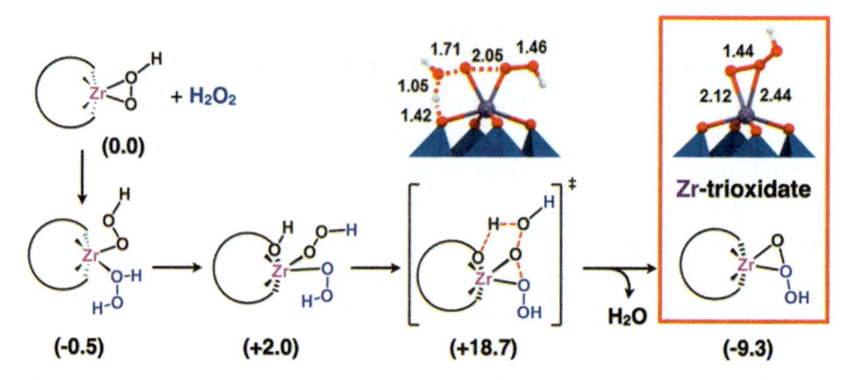

Scheme 6 Reaction mechanism for the formation of Zr-trioxidate, the key intermediate involved in the decomposition of H_2O_2. Relative Gibbs free energies are given in kcal/mol.

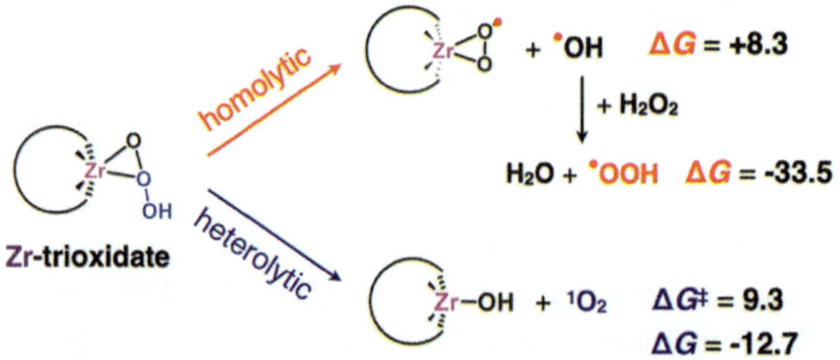

homolytic

+ •OH $\Delta G = +8.3$

+ H_2O_2

H_2O + •OOH $\Delta G = -33.5$

Zr-trioxidate

heterolytic

Zr–OH + $^1O_2$ $\Delta G^‡ = 9.3$

$\Delta G = -12.7$

Scheme 7 Decomposition of Zr-trioxidate via homolytic or heterolytic O–O cleavage pathways, yielding superoxide radicals and singlet oxygen, respectively. Reaction Gibbs free energies and free-energy barriers are given in kcal/mol.

This mechanistic proposal was further supported by experimental kinetic studies, which pointed to an intramolecular mechanism through a Ti-dihydroperoxide species and by comparison with other metal centers. Specifically, the experimental activation energies for H_2O_2 decomposition were determined to be 11.5, 14.6 and 16.7 kcal/mol for Zr(IV)-, Nb(V)- and Ti(IV)-containing Lindqvist anions, whereas DFT calculations predicted very similar zero-point-energy-corrected electronic barriers of 9.2, 14.7 and 18.6 kcal/mol, respectively *(70)*. Interestingly, the activation barrier increases on going from the Zr(IV)- to Ti(IV)-substituted POM, despite both metal centers having the same oxidation state, due to the inability of Ti(IV)-Lindqvist to activate the second H_2O_2 molecule through an inner-sphere mechanism. Furthermore, in going from Zr(IV) to Nb(V) within the same transition-metal row, the metal fragment becomes less nucleophilic and consequently, the nucleophilic attack of the Nb-hydroperoxo moiety on the second H_2O_2 molecule becomes less favored, increasing the apparent activation energy.

7. Outlook and perspectives

Since the publication of early studies in the 2000s, computational investigations have significantly contributed to broadening our understanding of POM-catalyzed epoxidation reactions with H_2O_2. The main findings distilled from the existing literature can be summarized as follows.

The reaction involves two main steps, as illustrated in Fig. 16: (i) the heterolytic activation of H_2O_2 on a TM-OH group to generate a metal-hydroperoxo or a protonated metal-peroxo complex (generally in equilibrium), in addition to a water molecule; (ii) the electrophilic oxygen transfer to the alkene substrate, yielding the epoxide product and regenerating the TM-OH catalyst. The second has been consistently found to represent the rate-determining step of the whole reaction, which can take place through several distinct mechanisms. Although protonated peroxo species are commonly more stable than hydroperoxo intermediates, the O-transfer step can take place from the latter through so-called "hydroperoxo paths". These can involve either the most electrophilic non-protonated alpha or the protonated beta oxygen. The alpha-oxygen transfer prevails in flexible and accessible metal centers such as Nb(V), Zr(IV) or under-coordinated and exposed Ti(IV) centers, while the beta-oxygen transfer, which involves a less distorted TS structure, occurs in more sterically-

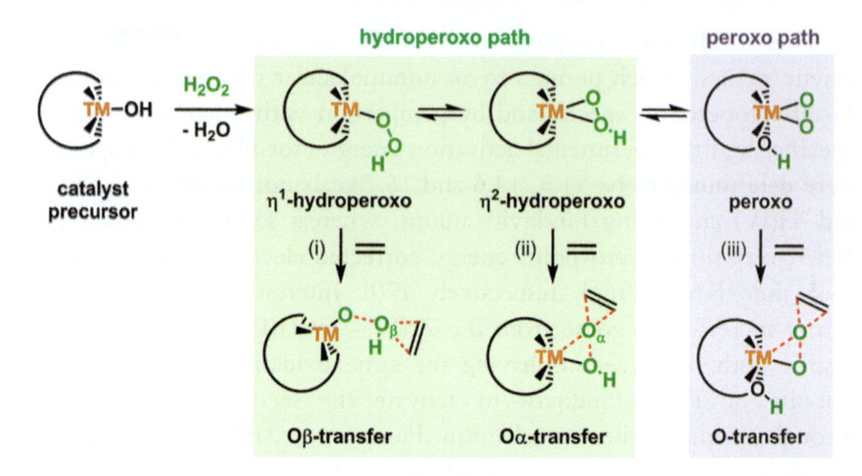

Fig. 16 Schematic overview of possible mechanisms for the epoxidation of alkenes with H_2O_2 catalyzed by POMs.

hindered metals or in those within a stiff coordination environment (hindered and coordinatively saturated Ti(IV) centers). The O-transfer step can also occur from the protonated peroxo intermediate through the "peroxo path" (Fig. 16), such as in divanadium-containing POMs in which the peroxo group can be cooperatively activated by two metal centers. Although somewhat controversial, evidences point also towards the "peroxo path" for lacunary W^{VI}-based catalysts. Additionally, in the presence of labile metal-oxygen bonds, such as in Ti-siloxy-POMs, alcohol-containing substrates that can coordinate the metal center can be also epoxidized through a hydroperoxo path in an inner-sphere fashion. This is favored over conventional outer-sphere paths thanks to a release in the strain around the active metal center granted by a partial detachment from the catalyst structure.

DFT calculations have also played a crucial role in elucidating the factors that influence the reactivity. In general, whatever factor that increases the electrophilicity of the active metal-hydroperoxo group (protonation of the POM, introduction of metals with high oxidation states, or using POM scaffolds with low negative charge) results in a decrease of the energy barrier for the rate-determining O-transfer step. Still, strong Lewis acids such as Nb(V) have been also found to over stabilize protonated peroxo resting states, being detrimental for the overall reaction rate. Similarly, while POMs with low negative charges may be expected to be beneficial for the reaction, one should note that the POM should be basic enough (that is, with a high enough negative charge) to associate protons to allow an efficient heterolytic activation of the H_2O_2 oxidant. Jointly, these findings underline that the electronic properties of the active metal centers and those of the usually non-innocent POM ligand have to be carefully adjusted to grant optimal catalytic performances.

Finally, recent efforts have been devoted to understand the competing unproductive H_2O_2 decomposition catalyzed by TM-POMs, which is the main side reaction that compromises the catalytic efficiency and product selectivity. The key step involves a nucleophilic attack of the peroxo intermediate to a second H_2O_2 molecule to form a metal-trioxidate species, which further evolves to yield the historically referred to "homolytic" products of H_2O_2 decomposition (involving $^1O_2$, $^{\bullet}OH$ and $O_2^{\bullet -}$). Thus, this implies that increasing the electrophilic character of the peroxo group does not only have a positive impact in accelerating the epoxidation reaction but also in slowing down the H_2O_2 decomposition side reaction, being beneficial for both activity and selectivity.

Overall, the gained fundamental knowledge in this catalytic process is profound and to date, the number of experimental and computational studies that focused on POM-catalyzed alkene epoxidation is significant. We believe that future efforts grounded on a strong dialogue between theoretical and experimental work hold the potential to guide the development of new generations of catalysts with enhanced or optimal performance, thus paving the way towards a greener and more sustainable oxidation catalysis framework.

References

1. Kholdeeva, O. A. *Eur. J. Inorg. Chem.* **2013,** *2013*, 1595–1605.
2. Smeets, V.; Gaigneaux, E. M.; Debecker, D. P. *ChemCatChem* **2022,** *14*, e202101132.
3. Kholdeeva, O. A. Selective Oxidations Catalyzed by Mesoporous Metal Silicates. In *Liquid Phase Oxidation via Heterogeneous Catalysis: Organic Synthesis and Industrial Applications;* Clerici, M. G., Kholdeeva, O. A., Eds.; John Wiley & Sons Inc: Chichester, 2013.
4. Kholdeeva, O. A.; Ivanchikova, I. D.; Maksimchuk, N. V.; Skobelev, I. Y. *Catal. Today* **2019,** *333*, 63–70.
5. Ivanchikova, I. D.; Zalomaeva, O. V.; Maksimchuk, N. V.; Stonkus, O. A.; Glazneva, T. S.; Chesalov, Y. A.; Shmakov, A. N.; Guidotti, M.; Kholdeeva, O. A. *Catalysts* **2022,** *12*, 742.
6. Dhakshinamoorthy, A.; Asiri, A. M.; Garcia, H. *Chem. Eur. J.* **2016,** *22*, 8012–8024.
7. Tombesi, A.; Pettinar, C. *Inorganics* **2021,** *9*, 81.
8. Nie, X.; Ren, X.; Ji, X.; Chen, Y.; Janik, M. J.; Guo, X.; Song, C. *J. Phys. Chem. B* **2019,** *123*, 7410–7423.
9. Wang, B.; Guo, Y.; Zhu, J.; Ma, J.; Qin, Q. *Coord. Chem. Rev.*, 476, **2023**, 214931.

10. Dong, J.; Zhu, H.; Xiang, Y.; Wang, Y.; An, P.; Gong, Y.; Liang, Y.; Qiu, L.; Zheng, A.; Peng, X.; Lin, M.; Xu, G.; Guo, Z.; Chen, D. *J. Phys. Chem. C* **2016,** *120*, 20114–20124.
11. Parker, W. O., Jr.; Millini, R. *J. Am. Chem. Soc.* **2006,** *128*, 1450–1451.
12. Gamba, A.; Tabacchi, G.; Fois, E. *J. Phys. Chem. A* **2009,** *113*, 15006–15015.
13. Spanó, E.; Tabacchi, G.; Gamba, A.; Fois, E. *J. Phys. Chem. B* **2006,** *110*, 21651–21661.
14. Gordon, C. P.; Engler, H.; Tragl, A. S.; Plodinec, M.; Lunkenbein, T.; Berkessel, A.; Teles, J. H.; Parvulescu, A.-N.; Copéret, C. *Nature* **2020,** *586*, 708–713.
15. a) (For recent examples, see) Bach, R. D.; Schlegel, H. B. *J. Phys. Chem. A* **2021,** *125*, 10541–10556; b) Bach, R. D.; Schlegel, H. B. *J. Phys. Chem. A* **2024,** *128*, 2072–2091.
16. Fraile, J. M. *Solid Catalysts for Epoxidation with Dilute Hydrogen Peroxide. Encyclopedia of Inorganic and Bioinorganic Chemistry;* John Wiley & Sons Inc,: Chichester, 2016; pp. 1–9.
17. Czapski, G. *Methods Enzymol.* **1984,** *105*, 209–215.
18. Hayyan, M.; Hashim, M. A.; AlNashef, I. M. *Chem. Rev.* **2016,** *116*, 3029–3085.
19. Kitajima, N.; Fukuzumi, S.-I.; Ono, Y. *J. Phys. Chem.* **1978,** *82*, 1505–1509.
20. Ono, Y.; Matsumura, T.; Kitajlma, N.; Fukuzumi, S.-I. *J. Phys. Chem.* **1977,** *81*, 1307–1311.
21. Khan, A. U.; Kasha, M. *Proc. Natl. Acad. Sci. U. S. A.* **1994,** *91*, 12365–12367.
22. Adam, W.; Kazakov, D. V.; Kazakov, V. P. *Chem. Rev.* **2005,** *105*, 3371–3387.
23. Haber, F.; Weiss, J. *Proc. R Soc. Lond [A]* **1934,** *147*, 332–351.
24. Weiss, J. *Trans. Faraday Soc.* **1935,** *31*, 1547–1557.
25. Kholdeeva, O. A. *Top. Catal.* **2006,** *40*, 229–243.
26. Pope, M. T. *Heteropoly and Isopoly Oxometalates;* Springer–Verlag,: New York, 1983.
27. Nymann, M. *Dalton Trans* **2011,** *40*, 8049–8058.
28. Simms, C.; Kondinski, A.; Parac-Vogt, T. N. *Eur. J. Inorg. Chem.* **2020,** *2020*, 2559–2572 (For a recent review, see).
29. López, X.; Carbó, J. J.; Bo, C.; Poblet, J. M. *Chem. Soc. Rev.* **2012,** *41*, 7537–7571.
30. Prabhakar, R.; Morokuma, K.; Hill, C. L.; Musaev, D. G. *Inorg. Chem.* **2006,** *45*, 5703–5709.
31. The "γ" notation stands for the isomerism of the Keggin anion. Specifically, in the γ isomer, two triads of edge-sharing tungstens of the Keggin anion are rotated by 60° with respect to their original arrangement in the α isomer. See López et al. (2012) for further details.
32. Nakagawa, Y.; Kamata, K.; Kotani, M.; Yamaguchi, K.; Mizuno, N. *Angew. Chem. Int. Ed.* **2005,** *44*, 5136–5141.
33. Kamata, K.; Yonehara, K.; Sumida, Y.; Yamahuchi, K.; Hikichi, S.; Mizuno, N. *Science* **2003,** *300*, 964–966.
34. Hussain, F.; Bassil, B. S.; Kortz, U.; Kholdeeva, O. A.; Timofeeva, M. N.; de Oliveira, P.; Keita, B.; Nadjo, L. *Chem. Eur. J.* **2007,** *13*, 4733–4742.
35. Musaev, D. G.; Morokuma, K.; Geletii, Y. V.; Hill, C. L. *Inorg. Chem.* **2004,** *43*, 7702–7708.
36. Kamata, K.; Hirano, T.; Kuzuya, S.; Mizuno, N. *J. Am. Chem. Soc.* **2009,** *131*, 6997–7004.
37. Kamata, K.; Ishimoto, R.; Hirano, T.; Kuzuya, S.; Uehara, K.; Mizuno, N. *Inorg. Chem.* **2010,** *49*, 2471–2478.
38. Ishimoto, R.; Kamata, K.; Mizuno, N. *Eur. J. Inorg. Chem.* **2013,** *2013*, 1943–1950.
39. Sartorel, A.; Carraro, M.; Bagno, A.; Scorrano, G.; Bonchio, M. *Angew. Chem. Int. Ed.* **2007,** *46*, 3255–3258.
40. Sartorel, A.; Carraro, M.; Bagno, A.; Scorrano, G.; Bonchio, M. *J. Phys. Org. Chem.* **2008,** *21*, 596–602.

41. Makrygenni, O.; Vanmairis, L.; Taourit, S.; Launay, F.; Sing, A. S. C.; Proust, A.; Gérard, H.; Villanneau, R. *Eur. J. Inorg. Chem.* **2020,** *2020*, 605–612.
42. Nakagawa, Y.; Kamata, K.; Kotani, M.; Yamaguchi, K.; Mizuno, N. *Angew. Chem. Int. Ed.* **2005,** *44*, 5136–5141.
43. Nakagawa, Y.; Mizuno, N. *Inorg. Chem.* **2007,** *46*, 1727–1736.
44. Kuznetsov, A. E.; Geletii, Y. V.; Hill, C. L.; Morokuma, K.; Musaev, D. G. *Inorg. Chem.* **2009,** *48*, 1871–1878.
45. Kamata, K.; Sugahara, K.; Yonehara, K.; Ishimoto, R.; Mizuno, N. *Chem. Eur. J.* **2011,** *17*, 7549–7559.
46. Kholdeeva, O. A.; Trubitsina, T. A.; Maksimovskaya, R. I.; Golovin, A. V.; Neiwert, W. A.; Kolesov, B. A.; López, X.; Poblet, J. M. *Inorg. Chem.* **2004,** *43*, 2284–2292.
47. Kholdeeva, O. A.; Maksimov, G. M.; Maksimovskaya, R. I.; Kovaleva, L. A.; Fedotov, M. A.; Grigoriev, V. A.; Hill, C. L. *Inorg. Chem.* **2000,** *39*, 3828–3837.
48. Antonova, N. S.; Carbó, J. J.; Kortz, U.; Kholdeeva, O. A.; Poblet, J. M. *J. Am. Chem. Soc.* **2010,** *132*, 7488–7497.
49. Wells, D. H.; Delgass, W. N.; Thomson, K. T. *J. Am. Chem. Soc.* **2004,** *126*, 2956–2962.
50. Wells, D. H.; Joshi, A. M.; Delgass, W. N.; Thomson, K. T. *J. Phys. Chem. B* **2006,** *110*, 14627–14639.
51. Skobelev, I. Y.; Zalomaeva, O. V.; Kholdeeva, O. A.; Poblet, J. M.; Carbó, J. J. *Chem. Eur. J.* **2015,** *21*, 14496–14506.
52. Donoeva, B. G.; Trubitsina, T. A.; Antonova, N. S.; Carbó, J. J.; Poblet, J. M.; Al-Kadamany, G.; Kortz, U.; Kholdeeva, O. A. *Eur. J. Inorg. Chem.* **2010,** *2010*, 5312–5317.
53. Kholdeeva, O. A.; Trubitsina, T. A.; Timofeeva, M. N.; Maksimov, G. M.; Maksimovskaya, R. I.; Rogov, V. A. *J. Mol. Catal. A Chem.* **2005,** *232*, 173–178.
54. Jiménez-Lozano, P.; Ivanchikova, I. D.; Kholdeeva, O. A.; Poblet, J. M.; Carbó, J. J. *Chem. Commun.* **2012,** *48*, 9266–9268.
55. Maksimchuk, N. V.; Maksimov, G. M.; Evtushok, V. Y.; Ivanchikova, I. D.; Chesalov, Y. A.; Maksimovskaya, R. I.; Kholdeeva, O. A.; Solé-Daura, A.; Poblet, J. M.; Carbó, J. J. *ACS Catal.* **2018,** *8*, 9722–9737.
56. Jiménez-Lozano, P.; Skobelev, I. Y.; Kholdeeva, O. A.; Poblet, J. M.; Carbó, J. J. *Inorg. Chem.* **2016,** *55*, 6080–6084.
57. Maksimchuk, N. V.; Ivanchikova, I. D.; Maksimov, G. M.; Eltsov, I. V.; Evtushok, V. Y.; Kholdeeva, O. A.; Lebbie, D.; Errington, R. J.; Solé-Daura, A.; Poblet, J. M.; Carbó, J. J. *ACS Catal.* **2019,** *9*, 6262–6275.
58. Zhang, T.; Mazaud, L.; Chamoreau, L.-M.; Paris, C.; Proust, A.; Guillemot, G. *ACS Catal.* **2018,** *8*, 2330–2342.
59. Clerici, M. G. *Kinet. Catal.* **2015,** *56*, 450–455.
60. Guillemot, G.; Matricardi, E.; Chamoreau, L.-M.; Thouvenot, R.; Proust, A. *ACS Catal.* **2015,** *5*, 7415–7423.
61. Zhang, T.; Solé-Daura, A.; Hostachy, S.; Blanchard, S.; Paris, C.; Li, Y.; Carbó, J. J.; Poblet, J. M.; Proust, A.; Guillemot, G. *J. Am. Chem. Soc.* **2018,** *140*, 14903–14914.
62. Solé-Daura, A.; Zhang, T.; Fouilloux, H.; Robert, C.; Thomas, C. M.; Chamoreau, L.-M.; Carbó, J. J.; Proust, A.; Guillemot, G.; Poblet, J. M. *ACS Catal.* **2020,** *10*, 4737–4750.
63. Zhang, T.; Solé-Daura, A.; Fouilloux, H.; Poblet, J. M.; Proust, A.; Carbó, J. J.; Guillemot, G. *ChemCatChem* **2021,** *13*, 1220–1229.
64. Zhu, B.; Lang, Z. L.; Yan, L. K.; Janjua, M. R. S. A.; Su, Z. M. *Int. J. Quantum Chem.* **2014,** *114*, 458–462.

65. Maksimchuk, N. V.; Evtushok, V. Y.; Zalomaeva, O. V.; Maksimov, G. M.; Ivanchikova, I. D.; Chesalov, Y. A.; Eltsov, I. V.; Abramov, P. A.; Glazneva, T. S.; Yanshole, V. V.; Kholdeeva, O. A.; Errington, R. J.; Solé-Daura, A.; Poblet, J. M.; Carbó, J. J. *ACS Catal.* **2021,** *11*, 10589–10603.
66. Jorda, E.; Tuel, A.; Teissier, R.; Kervennal, J. J. *Catal* **1998,** *175*, 93–107.
67. Yoon, C. W.; Hirsekorn, K. F.; Neidig, M. L.; Yang, X.; Don Tilley, T. *ACS Catal.* **2011,** *1*, 1665–1678.
68. Antcliff, K. L.; Murphy, D. M.; Griffiths, E.; Giamello, E. *Phys. Chem. Chem. Phys.* **2003,** *5*, 4306–4316.
69. Srinivas, D.; Manikandan, P.; Laha, S. C.; Kumar, R.; Ratnasamy, P. *J. Catal.* **2003,** *217*, 160–171.
70. Maksimchuk, N. V.; Puiggalí-Jou, J.; Zalomaeva, O. V.; Larionov, K. P.; Evtushok, V. Y.; Soshnikov, I. E.; Solé-Daura, A.; Kholdeeva, O. A.; Poblet, J. M.; Carbó, J. *J. ACS Catal.* **2023,** *13*, 10324–10339.

About the authors

Albert Solé-Daura earned his PhD from Rovira i Virgili University (URV, Spain) in 2020. Subsequently, he conducted postdoctoral research at the Collège de France (France), the University of Amsterdam (The Netherlands), the URV as a Margarita Salas fellow, and the Institut Català d'Investigació Química (ICIQ, Spain). Recently, in 2024, he was awarded a "la Caixa" Junior Leader fellowship to initiate his independent research career at ICIQ. His research interests concern the application of computational methods to delve into the dynamic properties and mechanisms of catalytic processes, primarily involving Polyoxometalates and Metal-Organic Frameworks.

Jorge J. Carbó obtained his PhD from Universitat Autònoma de Barcelona (UAB, Spain) in 2002. From 2002 to 2004, he did a postdoctoral stay at the University of Heidelberg (Germany). After a short postdoctoral stay in at the Institute of Chemical Research of Catalonia (ICIQ), he became a tenure track lecturer at the University Rovira i Virgili (Tarragona), in 2008 has was appointed as an associate professor, and in 2021 as a full professor. His research activity has focused on the application of computational methods to the study of molecular reactivity and catalysis including polyoxometalate chemistry, as well as, on the simulation of dynamic properties of nanosized molecular oxides.

Printed and bound by CPI Group (UK) Ltd, Croydon, CR0 4YY

02/12/2024

01798497-0013